Using Statistics in Social Rese

M000290084

Scott M. Lynch

Using Statistics in Social Research

A Concise Approach

Springer

Scott M. Lynch
Department of Sociology
Princeton University
Princeton, New Jersey, USA

ISBN 978-1-4939-5306-6 ISBN 978-1-4614-8573-5 (eBook)
DOI 10.1007/978-1-4614-8573-5
Springer New York Heidelberg Dordrecht London

© Springer Science+Business Media New York 2013
Softcover reprint of the hardcover 1st edition 2013
This work is subject to copyright. All rights are reserved by the Publisher, whether the whole or part of the material is concerned, specifically the rights of translation, reprinting, reuse of illustrations, recitation, broadcasting, reproduction on microfilms or in any other physical way, and transmission or information storage and retrieval, electronic adaptation, computer software, or by similar or dissimilar methodology now known or hereafter developed. Exempted from this legal reservation are brief excerpts in connection with reviews or scholarly analysis or material supplied specifically for the purpose of being entered and executed on a computer system, for exclusive use by the purchaser of the work. Duplication of this publication or parts thereof is permitted only under the provisions of the Copyright Law of the Publisher's location, in its current version, and permission for use must always be obtained from Springer. Permissions for use may be obtained through RightsLink at the Copyright Clearance Center. Violations are liable to prosecution under the respective Copyright Law.
The use of general descriptive names, registered names, trademarks, service marks, etc. in this publication does not imply, even in the absence of a specific statement, that such names are exempt from the relevant protective laws and regulations and therefore free for general use.
While the advice and information in this book are believed to be true and accurate at the date of publication, neither the authors nor the editors nor the publisher can accept any legal responsibility for any errors or omissions that may be made. The publisher makes no warranty, express or implied, with respect to the material contained herein.

Printed on acid-free paper

Springer is part of Springer Science+Business Media (www.springer.com)

For Mike and Betsy, my parents

Preface

This book began as a collection of lecture notes and has evolved over the last several years into its current form. My goal in putting the book together was to remedy some problems I perceive with extant introductory statistics texts. To that end, this book has several features that differentiate it from standard texts. First, typical introductory statistics texts are now commonly over 500 pages long. To be blunt, I don't believe that students actually read texts when they are so lengthy. So, although a 500+-page book may be chock-full of examples, illustrations, and interesting side information, such detail is offset by the fact that students will not read it. Here, I have tried to provide limited, but sufficient, examples of concepts and to keep the actual text limited to about 200 pages. In doing so, I have *not* covered a number of statistical methods that are not commonly used, like Fisher's exact test, Kruskal's Tau, and the Spearman correlation. While these, and similar, methods are certainly still used, they have limited application in contemporary social research. Thus, they can be investigated in detail by students when needed in the future (e.g., for a thesis or other research paper) via other sources.

Second, in this book, I integrate statistics into the research process, both philosophically and practically. From a philosophical perspective, I discuss the difficulties with concepts like "proof" and "causality" in a deductive scientific and statistical framework. From a practical perspective, I discuss how research papers are constructed, including how statistical material should be presented in tables and figures, and how results and discussion sections should be written.

Third, I emphasize probability theory and its importance to the process of inference to a greater extent than many books do. Most books discuss probability theory, but I feel its relationship with statistics is underdeveloped. In this book, probability is painstakingly shown to be the basis for statistical reasoning.

Fourth, I have included a number of homework problems at the end of almost every chapter, along with answers to the first half of them in each chapter in an appendix. There are several differences between the homework problems and answers I provide and those found in typical texts. First, I do not provide as many exercises as is commonly found in texts. In most chapters, there are approximately 10 exercises. The exceptions are Chap. 5, with 50 items, and Chap. 6, with 40 items. My rationale for this is that for developing an understanding of most concepts, I do not believe students need to perform dozens of exercises: once they have done one or two, especially ones that are word based and involve more thought in implementation, they understand the process. With problems involving more complicated (or at least tedious) calculations—like the chi-square test of independence, ANOVA, correlation, and simple regression—performing repetitive calculations may even be detrimental, fostering a "losing the forest for the trees" problem.

Thus, a second difference between my exercises and those often found in texts is that I have tried to construct exercises that are interesting, that often may be solved in different ways, that are realistic in the sense that they often require thinking like a full-fledged researcher would, and that often involve information that cannot simply be dropped into a formula in a rote fashion. For example, in the exercises for one-way ANOVA modeling, I have three exercises without solutions (and three with solutions). In one, the raw data are provided, and the student must compute sums of squares and complete the ANOVA table. This problem tests the student's ability to perform the basic computations. In another, a partial ANOVA table is provided, and all the student must do is understand the structure of the ANOVA table to complete it (i.e., $SST = SSB + SSW$ and $df(T) = df(B) + df(W)$) Finally, in another, the student is provided with group-level statistics like the mean and variance and must recover (two of) the sums of squares by "undoing" the variance calculations before constructing the ANOVA table. This problem tests the student's deeper understanding of the connection between variances and sums of squares.

A third difference between my exercises and those found in other texts is that I use the exercises in some cases to introduce additional material not covered directly in the chapter and as further examples of concepts and methods discussed in the chapter. As an example of the former strategy, the last exercise in Chap. 6 introduces the basic ideas underlying capture-recapture methodology. The detailed answers to exercises exemplify the latter strategy: the answers are detailed much as a solution to an example problem would be within a chapter in a typical text. They are placed at the end of the chapter so that they do not disrupt the flow of the text and so that students can work the problems and check their thought and computation processes as they work them.

A final difference between this text and others is that I routinely use simulated or contrived data, rather than real data, to illustrate concepts. To be sure, I use real data from the General Social Survey heavily in both examples and exercises. However, while real data are useful for giving students concrete examples, I think simulated data are often more useful when trying to illustrate difficult ideas, especially when showing how unknowns can be estimated. In real data, the unknowns are unknown! Also, it is generally the case that real data are messy. It is easier, for example, to illustrate the notion of row and column percentages in cross-tabs using rounder numbers than may be found in real data. Perhaps as a better example, it is easier to illustrate the idea of within versus between sums of squares in ANOVA modeling by creating contrived data in which the within sum of squares is 0 and contrasting it with data in which the between sum of squares is 0.

Layout for a Course

The book is intended as a text for a semester-long course on introductory statistics for social science students. Princeton semesters are relatively short (12 weeks), and our courses have one-day-per-week "precepts" in which the large class is broken into smaller groups (of about 15 students each) that are led by preceptors (graduate teaching assistants or the course instructor). I use these precepts to teach students how to use statistical software and to help students work through homework problems. Given the precepts, there are two 1-h lectures per week. After discounting 3 days for exams, that leaves 21 lectures. Table 1 shows the organization of the lectures and the corresponding book chapters.

As the table shows, I give four exams in my course. The first covers the basics of the research process and descriptive statistics. The midterm exam covers probability theory and one sample inference. The third exam covers various two-sample statistical tests, like the independent samples t test, and the chi-square test of independence. The final exam covers ANOVA, the correlation, and regression modeling.

Lecture	Topic	Chapter
1	Overview of research	1,2
2	Data acquisition	3
3	continued	
4	Summarizing data	4
5	continued	
*	First exam	none
6	Probability theory	5
7	continued	
8	Central Limit Theorem	6
9	One sample tests	
10	Confidence intervals	
11	Review	none
*	Mid-term exam	
12	Two-sample tests/intervals	
13	continued	
14	Chi-square	7
15	ANOVA	8
16	Correlation & simple regression	9
17	continued	
*	Third exam	none
18	Multiple regression	10
19	continued	
20	continued	
21	Tables and figures	11

Table 1. Typical course structure based on book

Acknowledgements

I have a number of people to thank for helping with the creation and revision of this book over the last 5 or 6 years. Most importantly, I thank my wife, Barbara, for listening to me talk about this project, and my undergraduate course more generally, over the last decade, for offering ideas for the book, and for being generally supportive. I thank my long-time colleague and friend Scott Brown for also providing numerous ideas for the book. I thank my colleagues, Mitch Duneier, for his help in revising the discussion of qualitative data and methods in Chap. 3, and Matt Salganik for providing general advice throughout the book. I thank my editor, Marc Strauss, and his assistant, Hannah Bracken, for their support in the publication process. Finally, I would like to thank the preceptors for my course over the last several years (since 2008), including Edward Berchick, Mariana Campos Horta, Phillip Connor, Elaine Enriquez, Lauren Gaydosh, Kerstin Gentsch, JoAnne Golan, Aaron Gottlieb, Angelina Grigoryeva, Patrick Ishizuka, Adam Moore, Manish Nag, Rourke O'Brien, David Pedulla, and Erik Vickstrom. Aside from the usual responsibilities of lecturing, grading, and meeting with students, these preceptors also helped find errors in the book and make suggestions for its improvement. Any errors that remain, however, are mine.

List of Symbols

$\rightarrow$	Logical if/then: $A \rightarrow B$ means "If A is true, then B is true"
$\therefore$	Logical/mathematical "therefore"
$\neg$	Negation (logical not) symbol. Ex: $p(A)+p(\neg A) = 1$ (the sum of the probabilities of event A and not event A is 1). Ex: $\neg(\neg A) = A$
$\equiv$	Equivalence: If $A \equiv B$, then B is just another way of expressing A
!	Factorial. $k! = k(k-1)(k-2)\ldots(k-(k-2))(1)(0!)$, with $0! = 1$ by definition
$C(n,x) \equiv \binom{n}{x}$	Combinatorial expression for selecting x items out of n items without replacement and in which order does not matter. Calculation: $\frac{n!}{x!(n-x)!}$
n	Commonly used to represent a sample size. N is often used to represent a population size
x_i	Generic representation of the ith value of the variable x. Example: If the variable x is body weight, x_3 would be the body weight of the third person in the sample under consideration
$\sum_{i=1}^{n}$	Summation of items labeled $i = 1 \ldots n$. i is the index of the summation loop; n is the total number of items being summed
$\bar{x}$	Sample mean
MD	Median
MO	Mode
IQR	Interquartile range
s^2	Sample variance; s is the standard deviation

$f(x)$	May be generic for a function of x, like a probability density function, or may represent the frequency of x
μ	Population mean
σ^2	Population variance; σ is the population standard deviation
s	Sample standard deviation
E_i	Generic event i that may occur in a trial
$p(E_i)$	Probability of event i occurring in a trial
S	Sample space, sometimes
$\in$	Is an element of. Ex. $S_1 \in S_2$ means that sample space S_1 is a subset of sample space S_2
$p(A, B) \equiv p(A \cap B)$	Joint probability of events A and B
$p(A \cup B)$	Probability of event A or B, where "or" is inclusive (and/or)
$p(A\|B)$	Conditional probability of event A *given* B is true
$P(n, x)$	Permutation of x items selected without replacement from n items in which order of selection/arrangement matters. Computed as: $\frac{n!}{(n-x)!}$
z	Usually a standard normal variable
p	Often the probability parameter in a binomial distribution. Can also be a "p" value, the probability of observing the sample data under the null hypothesis
e^x	The exponential function
$\sim$	"Is distributed as"
$\mu_{\bar{x}}$	The mean of the sampling distribution of means (i.e., the mean of the distribution of sample means under repeated sampling from the population)
$\sigma_{\bar{x}}$	The standard deviation of the sampling distribution of means. Under the Central Limit Theorem, $\sigma_{\bar{x}} = \sigma/\sqrt{n}$
$s.e.$	Standard error. Usually an estimate of $\sigma_{\bar{x}}$
H_0	Usually the null hypothesis
α	The value at which one is willing to reject the null hypothesis. If $p < \alpha$, we reject the null
t	A distribution similar to the normal distribution but with fatter tails and a degree of freedom parameter that, for our purposes, captures the extent to which s is a good estimate of σ

$t_{\alpha/2}$	The value of t for which one half of the mass α falls beyond in the t distribution
$1-\alpha$	The level of confidence in a confidence interval
$\min(a, b)$	Minimum function. Ex. $\min(1, 2) = 1$
χ^2	Chi-square. χ^2 is a probability distribution and is used in the chi-square test of independence and the lack-of-fit test in this book
$\bar{\bar{x}}$	The overall sample mean. Usually called the grand mean in ANOVA
R^2	R-squared. Proportion of variance explained in an ANOVA or regression model by grouping/predictor variables
MSE	Mean squared error from ANOVA and regression modeling
F	F is a probability distribution used in ANOVA and regression modeling
$cov(x, y)$	The covariance of x and y
r	Pearson's correlation coefficient
z_f	Fisher's z transformed correlation coefficient
$\approx$	Approximately equal to
$\hat{a}$	An estimate of a. Read: "a-hat"
α	Intercept term for the simple regression model. The estimate for it is $\hat{\alpha}$
β	Slope term for the simple regression model. The stimate for it is $\hat{\beta}$
e_i	Error term for the ith individual in the regression model
b_k	Alternate representation for regression coefficients. b_0 is the intercept, b_1 is the slope in simple regression, and subsequent subscripts are for additional variables in multiple regression models
σ_e^2	The MSE of the regression model. Usually has a "hat" above it, because it is estimated and not directly observed

Contents

List of Figures

List of Tables

Chapter 1
Introduction

Statistics plays a crucial role in scientific research, especially social science research based on observational (nonexperimental, survey) data. In addition, statistics plays an increasingly important role in our daily lives. We are constantly inundated with statistics in the news in the form of political polls and general information about one or another aspect of life, and these statistics routinely find their way into politics—often as sound bites—as well as into policy.

At the same time that the use of statistics permeates our lives, statistics used as a evidence for a position or argument are much maligned, and statistics as a course of study is dreaded. Few college students look forward to taking a statistics course, few college graduates seem to recall their introductory statistics course fondly, and in the general population, people never seem to have anything good to say about the subject. Indeed, a number of colloquialisms have emerged about statistics, including: "there are lies, damned lies, and statistics," "you can prove anything with statistics," and "87 % of statistics are made up on the spot" (this percentage, of course, can be replaced with any number you like!).

In fact, "statistics" don't prove anything by themselves; they must be interpreted. For example, a 10 % unemployment rate in the US could reflect any number of things. It could mean that the country has a lot of lazy people. It could mean that the economy is in a recession, with businesses cutting payroll in order to maintain a stable level of profit. It could mean that there is no recession, but simply that the working-age population is growing faster than demand for products and services. In short, such a statistic in the absence of other data could mean just about anything. And using such a statistic as "proof" of some point is no more valid than using poor logical reasoning with words is. Put another way, one could just as easily—and falsely—say that "you can prove anything with words." The difference is simply that statistics, because their production involves mathematics—another area of study often dreaded—have an air of mysteriousness and sophistication about them that seems to lend greater legitimacy to arguments that involve them than those that do not.

This is not to say that statistics are useless or that learning about statistics is a waste of time. There is a right way (or right ways) to use statistics, and the

S.M. Lynch, *Using Statistics in Social Research: A Concise Approach*,
DOI 10.1007/978-1-4614-8573-5_1, © Springer Science+Business Media New York 2013

main goal of this book is to show how to use statistics properly in social science research. A secondary goal is to make you aware of how statistics can be misused, misinterpreted, or faultily constructed in a variety of settings, especially in the media and in social and political debate.

As you will learn in the book, there are three main purposes of statistics: summarizing data, making inference from samples to populations, and predicting (or "forecasting") future—or at least unobserved—events based on extant data. Most people are aware that statistics are used to summarize data. For example, when a news agency reports that 15 % of the US population does not have health insurance, this is a single number summarizing the status of a population of 300 million people.

Fewer people are aware that this type of percentage is not usually based on a complete enumeration and canvassing of the population, but rather a small sample drawn from it. Drawing conclusions about the entire population from a small sample is the purview of "inferential statistics," and this aspect of statistics is less well understood. Indeed, when I read viewer comments on news websites reporting the results of polls measuring political attitudes, I often see comments questioning the validity of the results based on several recurring arguments. Some question how a poll of 600 persons in the US could possibly accurately reflect the attitudes of 300 million people. Some suggest that poll results are inherently flawed because "I've never been polled, and my views (and those of everyone I know) disagree with the results." As you will learn in this book, extremely precise and accurate results—"inferences" to the larger population (in fact, an infinitely large population)—can be obtained from very small samples. Furthermore, given the small samples that are usually taken in polls, it is not surprising that most persons will never be involved in a "scientific" poll. Indeed, a very rough calculation suggests that at least a half million polls would have to be taken before any given individual in the U.S. population could expect to be polled. While polls have become increasingly common over the last few decades, far fewer than 500,000 polls have been conducted.

In terms of prediction, we often hear statements like: by 2037, the Social Security system will be bankrupt, or, by 2020, white non-Hispanics will no longer be the majority of the US population. These sorts of claims also fall within the purview of statistics. Such claims are based on projecting the current state of the world forward in time based on knowledge of observed patterns of fertility, mortality, migration, tax rates, etc. We often refer to such predictions as "forecasts," because their predictive validity—that is, their ability to accurately predict future events—rests on the assumption that future conditions and patterns will remain as they have been, or at least as they are expected to be. If there is a change from expectation, then all bets are off. For example, imagine the consequence for solvency if Social Security tax rates were to increase (or decrease).

Another type of prediction in statistics regards the ability to make statements about unobserved situations or individuals based on observed situations or individuals. For example, suppose I observe in a sample that men, on average, earn more than women, on average. I might then predict that, in a future random sample, the men will earn more than the women on average. Many people have trouble with this type of statistical argument. Some might say that they know plenty of women who earn more than men that they know—and this may be true. For example, in the 2008 General Social Survey (data from which will be used throughout this book, and so it will be abbreviated as GSS; see Smith et al. 2011), the average income for men in the sample was just over $48,000, while the average for women was just over $32,000. However, there were certainly many women in the sample who earned more than some men (and vice versa). In particular, 20 % of women in the sample earned more than the average for men, and 36 % of men earned less than the average for women.

In other words, both statements are true; one does not contradict the other. To argue that the claim about average income is false because of one's own experience is to ignore the fact that one's own experience may not be representative of all of reality, while the average is a summary of all of reality. Put another way: individual experience does not invalidate a general pattern, so long as that general pattern is properly established. We call such appeals to one's personal experience "anecdotes," and it does not matter how many anecdotes one can provide, they do not establish or reject patterns established by actual data.

A more substantial argument may be that my claim about average income perpetuates an invalid "stereotype" about women. This argument is somewhat more difficult to counter, and this is one dilemma that anyone using statistics inevitably will come across. By definition, a stereotype is a generalization about a "type" of individual that has a "kernel" of truth. In the example just discussed, the fact is that there are more women who make less than the average income for men than there are above the average for men. So, my claim is not a stereotype: it is more than a "kernel" of truth. Think about it this way. If I selected a man and woman at random from the population, would you be willing to place a significant bet that the woman earns more than the man? Given that the average tells us something about where the distribution of incomes is centered for men and women, it would probably not be a smart bet.

Importantly, one thing that differentiates a statistical argument from a stereotype is that a statistical argument generally places limits on its claim. For example, suppose it's true that unemployed persons are more likely to engage in robbery than employed persons, such that 2 % of the unemployed commit robberies, while 1 % of the employed commit roberies. While it's true that the unemployed commit robberies at twice the rate of the employed (the "kernel of truth" here), it would be fallacious (and ridiculous) to fear the unemployed but not fear the employed. In neither group is robbery a predominant behavior, and so to claim that unemployed persons are robbers would be to unjustly "stereotype" the unemployed as criminal.

This type of stereotypical reasoning may be obvious, but there are certainly less obvious instances of it. For example, in reading viewers' comments on news websites, I have often seen the argument that all unemployed persons are simply lazy, bolstered by the argment that the commenter knows one or more lazy unemployed persons. The claim that ALL unemployed folks are therefore lazy is fallacious, because it is a "hasty generalization." The generalization is "hasty," because a single person's view of his/her own experience (observing, usually, only a couple of persons at most) is not representative of the population. The opposite argument—that, because "I know one unemployed persons who is not lazy, therefore no unemployed persons are lazy"—is equally fallacious. Therefore, debates involving this type of faulty reasoning can go nowhere. Instead, properly used statistics may help arbitrate the issue.

In short, we cannot generalize—in any direction—from our personal experience. This is one of the most difficult concepts to learn in statistics, and yet it is one of the most important. To be blunt, be wary of any claim that says that "all X are Y." Sound statistical reasoning involves making generalizable claims, but equally importantly, it places qualifiers and limits on those claims.

1.1 Goals of This Book

In this book, we will focus on the role of statistical reasoning and methods in social science research, paying particular attention to the roles of summarization and inference. As such, the important starting point is discussing what scientific research is about—as well as what it is not about. The next chapter discusses the process of scientific research, from the stage of asking a general question through the specification of hypotheses that are to be tested using statistics. Chapter 3 discusses how to find and/or collect data to address a hypothesis, with a particular focus on issues of sampling design and construction of quantitative survey instruments. Although this book is not geared toward showing you how to develop your own instrument and collect data—indeed, the focus of the book is on using extant ("secondary") data—it is extremely important to understand how one's data was collected, and how survey questions were asked, in order to produce meaningful statistical analyses.

Chapter 4 shows how to take a data set and produce meaningful summaries of it using descriptive statistical methods. The chapter begins by discussing how to summarize univariate data, that is, how to summarize single variables that may be of interest, like years of schooling or earnings, before showing how to summarize bivariate data using "cross-tabulations."

Chapter 5 discusses probability theory in considerable depth in order to lay the foundation for understanding statistical inference. Chapter 6 begins by introducing one of the most difficult ideas in introductory statistics that follows from probability theory—the Central Limit Theorem. The chapter then shows how statistical inference can be made using the theorem to conduct "hypothesis tests" and to construct "confidence intervals" for univariate data, that is, data consisting of a single variable. The chapter continues this discussion in extending inference about one variable to inference about group differences in a continuous variable.

Chapters 7 through 10 extend the basic concepts of inferential statistical reasoning for single, continuous variables, to different types of quantitative data and more sophisticated types of research questions.

Finally, Chapter 11 shows how to summarize the results of statistical analyses in reader-friendly tables and figures, as well as how to summarize the results and write the discussion and conclusion sections of a scientific research paper.

Chapter 2
Overview of the Research Process

Scientific research is the process of (1) developing an empirically answerable question, (2) deriving a *falsifiable* hypothesis derived from a theory that purports to answer the question, (3) collecting (or finding) and analyzing empirical data to test the hypothesis, (4) rejecting or failing to reject the hypothesis, and (5) relating the results of the analyses back to the theory from which the question was drawn. This last step usually involves revising the original theory to handle discrepancies between what the empirical data show and what the original theory posited, although the findings of only a single study usually are not sufficient to warrant major revisions of a theory. Nonetheless, a scientific research study, no matter how small the contribution, *must* make a new contribution to be considered original scientific research. In other words, a research study adds to our knowledge base. This requirement distinguishes a research paper from a report.

For the most part, the scientific understanding of a topic changes slowly, in part because science is about falsifying theories—and not proving them—and in part because a well-done study can generally only address a single, very narrow research question or hypothesis. Occasionally, some research leads to major changes in theory or a major change in an entire orientation that gives rise to theory (a "paradigm shift"; see Kuhn 1962), but generally, science progresses very slowly, with theories being "fine-tuned" with each additional study.

2.1 What Is NOT Research

Based on the above definition of scientific research, there are at least three things we can say do *not* constitute scientific research. I discuss these here, because they are commonly misconceived as constituting research, but they are not. The remainder of this book is geared toward discussing the role of statistics in actual research, and not in report writing.

S.M. Lynch, *Using Statistics in Social Research: A Concise Approach*,
DOI 10.1007/978-1-4614-8573-5_2, © Springer Science+Business Media New York 2013

1. A literature review alone is not research.

   Although reviewing and synthesizing literature is an important part of the research process, it is not, in-and-of-itself, research. The process of reviewing literature does not add to our knowledge base. In the "hard sciences," for example, writing a paper that delineates the history of Newtonian physics and the emergence of Einsteinian quantum physics does not constitute research. Any physicist knows this history without engaging in any experimental or nonexperimental work. Similarly, in social science research, writing a review of the state of the literature on race differences in educational attainment, for example, does not constitute original research.
2. Theory construction or forging links between theories or perspectives in the absence of empirical data is not research.

   Although the process of modifying and perhaps linking theories is an important part of research, it is not research by itself. For example, forging links between Newtonian physics and Einsteinian physics is not research. Any physicist should be able to derive such links analytically or at least conceptually. Similarly, in sociology, relationships between Durkheim's concept of anomie (see Durkheim 1997) and Marx's concept of alienation (see Marx 1988) are derivable without reference to any empirical data. In other words, theory construction and linking theories are logical, but not empirical processes. Furthermore, while it may contribute to our understanding of anomie and alienation, it does not contribute anything new to our understanding of the world.
3. Collection and analysis of data without theoretical grounding (i.e., having no research question) and offering no explanation of the data is not research.

   Data collection is an incredibly important—central—part of research; without data, no research could be done. However, data collection alone is not research, because there is no theory-driven question being asked nor answered. For example, the Census Bureau collects data on the size of the US population every decade (as well as additional factors, like age structure, racial composition, etc.), and it usually produces numerous reports on the various features of the population. This data collection and reporting is essential to social science research, because social scientists often use Census data in their own research, but the Census Bureau's collection and reporting process itself is not research, because there is no research question driving the collection and reporting. In other word, there is no direct contribution to our knowledge base regarding a topic of scientific interest.

2.2 Replication as Research

One activity that may not appear to be research, but in fact is, is replication. Replication is the process of repeating an experiment or study to verify the original findings. Technically speaking, true replication involves mimicking a previous study exactly; that is, using the same data collection and analysis method, and observing whether the original research results hold. As a simple example of replication, take the construction of a volcano for a science fair project in junior high school. The typical such exercise involves building something that looks somewhat like a volcano (usually, this is where students concentrate most of their effort, although this is not the actual "science" part of the process) and mixing vinegar and baking soda in a container at the volcano's peak. Mixing these two substances produces a chemical reaction that causes "lava" to spew out of the top and pour down the sides. If the student hypothesized that the mixture of an acid (vinegar) with a base (baking soda) will produce a chemical reaction that generates a gas and something else, and the student verified this with the empirical data (the "eruption"), then this process could be considered replication of a test of some centuries-old chemistry hypothesis.

Replication is an important part of the scientific process, although it receives very little attention in contemporary social science literature. Frankly, it is very difficult to get such work published. Usually, seemingly replicative research is not truly replicative; instead, it may differ in data and method very slightly from the original work. For example, it may involve analyzing data from a previous study in a slightly different way. Technically, this does not constitute replication, but it is research. More importantly, this type of work constitutes the bulk of social science research: most social science research elaborates on extant research by expanding it in relatively minor ways. For example, one study may find that stress is related to mortality risk in the general population. A second study—using the same data, perhaps—may find that this relationship holds for persons over age 45, but not for persons under 45. This finding then leads to a revision of the theory that produced the hypothesis that stress affects mortality, but it does not invalidate it—it simply qualifies it.

2.3 Stages of Research and Scientific Paper Structure

Although I have defined the overall process of research as involving five primary steps, the process of research can be decomposed more specifically into a number of sub-steps. Table 2.1 shows this decomposition and its relationship to the parts of scientific research papers seen in contemporary research journals.

These ten steps in the research process can be subdivided into additional steps, and some of these steps will be the focus of the material covered in this book. We will only briefly review the first five steps—corresponding to the introduction and theory sections of a research paper—in this chapter; steps 6–10 will constitute the focus of the remainder of the book. I note at the outset that these research

The Research Process	Paper Section
(1) Start with a Perspective (2) Select a Theory (3) Derive a Proposition (4) Ask a Question (5) Derive Hypotheses	Introduction & Lit. Review/Theory
(6) Find or collect data (7) Analyze data	Data & Methods
(8) Report results & answer question	Results
(9) Interpret results in terms of theory (10) Draw implications for theory	Discussion & Conclusions

Table 2.1. Correspondence of the parts of the research process with the parts of a scientific paper.

components as I have spelled-them out may not be universally agreed-upon. For example, there are several ways to define a proposition and a hypothesis. As another example, I have seen some definitions of a proposition and a research question that equate them; their only difference in that case being whether one poses it as a question (research question) or a statement (proposition).

2.3.1 What Is a Perspective?

A perspective is a general orientation toward the world. Perspectives are ultimately untestable but simply frame the world in a particular way. Therefore, there really shouldn't be much room for serious debate about them, and research ultimately can't confirm nor disconfirm them. For example: are you politically liberal or conservative? Do you have a generally optimistic or pessimistic view of the world? Do you view human nature as inherently good or inherently evil?

As a more specific example of perspectives, within the social sciences (sociology in particular) the two chief perspectives are the functionalist and conflict perspectives. The functionalist perspective views the social world as a system of components that function together to maintain the whole. The conflict perspective views the social world as a system of antagonistic components held together (perhaps) by force. From the functionalist perspective, a behavior like crime can be viewed either as (a) a dysfunction that is corrected or held in check over time via institutions like the criminal justice system, or (b) a phenomenon that actually serves a function—like demarcating and clarifying boundaries between acceptable and unacceptable behaviors when it is punished. A conflict perspective, on the other hand, might view crime as a response to the oppression of one social class (the poor) by another (the wealthy), or it might view it as one more way the wealthy control the poor (e.g., consider the punishment for blue collar versus white collar crime).

These larger orientations toward the world can not be confirmed nor disconfirmed by data for several reasons. First, they are simply too broad: even if one *could* find evidence that supported, say, a contemporary liberal position on a given topic, the evidence could not be construed as validating liberalism as a whole. For example, there is little doubt that Social Security—a liberal policy—improved the quality of life for elders in the US. That fact, however, does not imply that the government always has the best (or even a) solution to any social problem.

Second, and relatedly, general perspectives may seem right in some contexts and wrong in others. Consider, for example, the statement that "the government should stay out of our lives." This view is held by both political liberals and conservatives, but in different contexts. When liberals make this statement, they mean that the government should not be able to restrict behaviors, like recreational drug use and homosexual marriage. When conservatives make this statement, they mean that the government should not be able to levy heavy taxes or regulate businesses. Very few hold views that are both socially and economically liberal (or conservative).

Third, and perhaps most problematic, almost any empirical data that are observed can be interpreted as being consistent with any perspective. In general, perspectives are so broad that they have more than one way—and possibly contradictory ways—to predict or explain any data. For example, suppose I believe that humans are selfish by nature. I might predict, then, that no one would jump in front of a bus, risking certain death, to save a child from being hit. Yet, certainly this sort of behavior happens, and it does fairly often. Does this occurrence invalidate my perspective? No: there could be numerous followup explanations that make the data consistent with—or irrelevant to—the perspective. One might argue that the individual was ultimately being selfish, because s/he simply wanted recognition for being a hero and possibly miscalculated the risk. One might argue that the individual may have been temporarily insane, not making a rational calculation at all (either selfish or altruistic). One could argue—if the person the individual saved were a relative—that the act was a selfish attempt to further the family genes. If the person were unrelated, one could argue that saving the child helps preserve the species for at least another generation.

In short, a perspective cannot be rejected. Data can either be interpreted directly to support the perspective, or a question-begging (i.e., circular) argument can be made to oppose evidence to the contrary: Altruistic behavior cannot occur, because I can interpret any behavior that seems altruistic in a fashion that indicates it isn't.

If perspectives cannot be tested, then why are we discussing them here as part of the scientific process? I list starting with a perspective as part of the process, because we need to be aware that we all start with particular orientations toward the world. We are not *tabula rasa,* that is, blank slates that simply digest data as it emerges. Instead, we all view the world through particular lenses, and these lenses affect how we interpret data. Put another way, data are not unambiguous; they do not "speak for themselves." To guard against letting our own biases get the better of us while we are conducting research, we need to first recognize that we have them. One way to reduce the influence of our biases is to work at a "lower" level than that of a perspective; that is, to derive smaller, more manageable, testable (and falsifiable) ideas.

2.3.2 *What Is a Theory?*

A theory is a systematic description of how the world (or part of it) "works." Generally, a theory is too broad to be tested in its entirety in a single study, but it offers a framework for understanding how things in the world operate. Therefore, theory enables prediction, which we will discuss shortly (as hypotheses). Within sociology, Durkheim's theory of modernization seeks to explain why modern societies "hold together" as a unit rather than fragment rapidly into anarchy or chaos, and what the consequences of maintaining a society are (Durkheim 1997).

A short version of this theory is as follows. Pre-modern societies maintain stability because people within them are very similar to one another—everyone knows how to (and does) perform the basic tasks necessary for survival, everyone shares similar understandings of events like why the sun comes up, why it rains, etc. With such similarities, there is little to disagree about, and so the society is held together by a force that Durkheim called "mechanical solidarity," where "mechanical" refers to the automatous nature of people in the society. However, as societies develop the ability to produce surpluses (especially of food, e.g., via agricultural revolutions), a division of labor begins to emerge. That is, people have the ability to specialize in particular types of labor once the basic necessities of life for all of a society can be produced by a few (e.g., in the US, approximately 2 % of the population is engaged directly in food production; thus, the other 98 % can do something else). However, with the division of labor come two significant changes. First, people become interdependent; they are no longer self-sufficient. Second, the commonality of societal beliefs begins to fragment. Regarding the first change—the emergence of interdependence—Durkheim argued that societies still hold together, but the "glue" changes. Society no longer evidences mechanical solidarity, but rather "organic solidarity," where organic solidarity is the interdependence of the components of society (think of the organ systems of the body—each has its own function, and none can exist without the others). The second change—the erosion of common beliefs—stems from the dissimilarity in the day-to-day lives of the members of the society. For example, consider the different lives of a scientist and a preacher, and consider the different sets of beliefs held by the two (e.g., consider evolution vs. creationism). Eventually, Durkheim argued that the dissimilarity of the members of a society would lead to "anomie," a state of normlessness in which society's members would be uncertain about how to behave. At the individual level, this condition of anomie would produce angst.

Notice how this theory is fairly broad and encompasses a number of steps. No single research project could possibly address all the components of the theory.

2.3.3 *What Is a Proposition?*

A proposition is a *single* potentially testable component of a theory. The theory described above is a condensed version of Durkheim's original theory, but yet a number of propositions can be derived from it. One proposition, for example, might be that societal development of a division of labor generates anomie.

Notice that this proposition follows from the original theory, but it does not encapsulate ALL that the theory entails. It does not, for example, say anything about any prior stage of the theory, in terms what leads to the production or expansion of a division of labor.

2.3.4 *What Is a Research Question?*

A research question may be considered a refined restatement of a proposition that is testable. For example, we can revise the proposition above as: Do societies with more extensive divisions of labor evidence greater levels of anomie? (or: Does the expansion of a division of labor produce anomie?) Notice that the research question is (1) framed as a question, and (2) makes the proposition testable by suggesting a way to evaluate it: societies may be compared on measures of division of labor and anomie in order to answer the question.

2.3.5 *What Is a Hypothesis?*

A hypothesis is a *falsifiable* statement that makes a prediction derived from the research question that suggests a relationship between variables. In other words, it is a statement *that can be disproven.* A single research question may generate several hypotheses. The refinement of a research question into a hypothesis typically includes clarifying exactly what some of the concepts in the proposition and question mean. For example, what is a division of labor? What is anomie? How do these concepts link with empirically-measureable phenomena?

Some methodologists say that hypotheses are statements about expected relationships between variables that "operationalize" (i.e., measure) the concepts discussed in the research question. Often, however, the hypotheses presented in scientific articles are presented in the second section of the paper (the literature review and theory section), while the actual operationalization of the concepts occurs in the data section. Regardless, hypotheses stem from the research question and are more specific. So, for example, regarding the research question above, a hypothesis might be: societies with a more heterogeneous occupational structure will have members with higher levels of anxiety. Here, we have refined the terms "division of labor" and "anomie" with the terms "heterogenous occupational structure" and "anxiety." We certainly need to clarify these terms as well, but this is generally done in the data section.

2.4 Some Summary and Clarification

The distinction between theory, propositions, research questions, and hypotheses is admittedly somewhat fuzzy. This is largely because there are different definitions for each across generations of researchers and across scientific disciplines. Within social science research there has been a trend over the last half-century away from the formal spelling-out of propositions and hypotheses in research articles. Instead, most articles now simply present a research question, followed by theory that purports to answer the question, followed by a test of part of the theory. Even the magnitude of a theory may differ across fields and time. For example, early work in sociology was concerned with developing "grand theories" that were large and attempted to explain all of human behavior. Over the last several decades, however, smaller "middle range" theories have emerged to explain more narrow forms of behavior (Merton 1968). These are often called "substantive theories" today.

The location of the research question as the fourth step in the research process is also tenuous. We generally begin with a question of interest (perhaps call this step 0). However, as we learn more about research in the topic area, more often than not, we find that our initial question is too broad and has to be refined. Thus, one could consider our original question as a "research question," but, based on our outline here, the true "research question" will be much narrower and more precise than our original question.

2.5 What Research Cannot Do: Proof

The process of moving from a theory to a proposition to a research question and hypotheses is a *deductive* one, that is, each stage follows from (is a *necessary consequence of*) the stage prior to it. A consequence of this deductive process is

Symbolic	Generic language	Example
$A \rightarrow B$	If A is true, then B is true	If it rains today at my home, my yard will be wet
B	B is true	My yard is wet
$\therefore$	Therefore,	Therefore,
A	A is true	It rained

Table 2.2. An example of the fallacy of affirming the consequent.

that theories cannot be proven. Put another way, *science is not a process of proving theories.* Some basic logic shows the problem with the notion of proof. Consider the argument structure presented in Table 2.2.

At first glance, the argument may seem reasonable. However, it is invalid—it is a "fallacy of affirming the consequent"—because it is possible that $C \rightarrow B$. In other words, B may be a consequence of A, but it may also be the consequence of some other premise (or theory), C. It is possible, for example, that 1,000 (or more) leprechauns urinated on my lawn while I was away. More realistically, my wife may have watered the lawn. In short, the data may be perfectly consistent with my original theory but data cannot, in fact, validate it.

More generally, if a theory makes a prediction regarding some pattern in data, the fact that real data may evidence the predicted pattern does not prove the theory true, because another theory may make the same prediction. To make this idea more concrete, suppose the data do in fact show that societies with heterogenous occupational structures have higher rates of anxiety. This does not suggest that Durkheim's theory is correct, because an alternate explanation might be Marx's theory of alienation resulting from capitalistic exploitation of labor (Marx 1988). As another example, in physics, most predictions about the physical world were consistent with classical, Newtonian physics. However, once better measurement capabilities emerged, physical reality was more consistent with quantum theory, leading to a shift in support for that theory. In sum, when we find evidence that is consistent with a hypothesis/theory, we can only say that the evidence is consistent with the hypothesis/theory. The evidence does not prove the theory true (see, for example, Hawking 1988).

Although we cannot prove a theory true through empirical research, we may be able to disprove one. Consider an alternative, but similar logical structure, as shown in Table 2.3. Although this argument structure looks like the one presented before, it is fundamentally different. It says that, if we find out that the consequent (B) of a conditional statement ($A \rightarrow B$) is untrue, then the original premise (A) cannot be true. In other words, if a theory predicts some pattern will be observed in empirical data, and that pattern is not observed, then the theory cannot be true. For that matter, no theory that makes the same prediction could be true. This argument structure is

Symbolic	Generic language	Example
$A \rightarrow B$	If A is true, then B is true	If it rains today at my home, my yard will be wet
$\neg B$	*B* is not true	My yard is not wet
$\therefore$	Therefore,	Therefore,
$\neg A$	*A* is not true	It did not rain

Table 2.3. An example of the logically valid argument structure of *modus tollens*.

logically valid and is called "modus tollens" (see Bonevac 2003). Scientific research is therefore geared toward falsifying theories—or hypotheses drawn from them—and not proving them true (see Popper 1992).

Given that the goal of science is to falsify theories, there is a considerable amount of confusion between the scientific conception of a theory and the common usage of the term. Consider, for example, the ongoing debate regarding evolutionary theory, in which opponents have denigrated evolution by claiming it is merely "a theory." In contrast to the common view of a theory as mere speculation, a scientific theory is *not* simply speculation. It is an explanation for how some aspect of the world works that is falsifiable and has "stood the test of time" by not being falsified by evidence in repeated studies (Coyne 2009). Given that it is logically impossible to prove a theory true, there is no greater status for an explanation to attain than as a "theory."

2.6 Conclusions

In this and the previous chapter, we discussed the process of scientific investigation and its relationship to the components of a research paper. This process—of science and scientific writing—is more or less the same for every discipline, from the so-called "hard" sciences to the social sciences. The key difference across disciplines is in the subject matter, and therefore, the type of data used and the methods for gathering it. We will discuss these issues in the next chapter before turning to the process of analyzing data in Chap. 4 and beyond.

2.7 Items for Review

Be familiar with the following concepts, terms, and items discussed in this and the previous chapter:

- Falsification/falsifiability
- Stereotype

- Hasty generalization
- Replication
- Scientific theory
- Proposition
- Research question
- Hypothesis
- Deduction
- Fallacy of affirming the consequent
- *Modus tollens*

2.8 Homework

Answer the following questions:

1. What differentiates the scientific definition of a theory from the common-use definition of the term?
2. Why can a theory not be proven?
3. Why is a perspective untestable?
4. Explain the difference between a research question and a hypothesis.
5. What does "falsifiable" mean?
6. Give three examples of the fallacy of affirming the consequent.
7. Give three examples of arguments that use *modus tollens.*
8. Explain how the *modus tollens* argument form is useful for science.

Chapter 3
Data and Its Acquisition

Once a research question has been established and some hypotheses have been derived, the next stage in the research process is to determine what type of data is needed to answer the question/hypotheses. There are two basic types of data in social science research: quantitative and qualitative. Quantitative data is data that represents items of interest numerically, and quantitative research involves examining patterns in such data using statistical methods. Examples of quantitative data include height measured in inches, IQ scores, years of schooling, earnings, counts of depressive symptoms, measures of attitudes, etc. Qualitative data represents small numbers of cases—situations, experiences, events—using data from observations, interviews, or archives that are usually not chosen using probabilistic methods. The phenomena investigated usually cannot be fully understood via quantification. For example, what is the process of death like for a dying person? How do caregivers deal with the death of a loved one who has suffered tremendously before death? What is it like to participate in an illegal activity like dog fighting? What is the life of street vendors in NYC like? Qualitative research involves examining responses to these types of questions interpretatively for common themes in order to understand human experience, often in marginal populations. A key distinction between quantitative and qualitative approaches is that much quantitative research is oriented toward making inferences about causal processes, while qualitative research is not.

Almost every broad topic can be studied qualitatively or quantitatively. For example, some research in stratification—traditionally a quantitatively-dominated area of study—has focused on the experience of being poor. A quantitative study may involve surveying a large number of people probabilistically to determine the effect of growing up in a poor neighborhood on developing childhood obesity. In contrast, a qualitative study might involve in-depth interviews with a few individuals in a poor neighborhood and ask what it is like to live in a neighborhood where there are no fresh fruits or vegetables available. Over the past decade or so, research involving *both* qualitative and quantitative data collection and analysis (called mixed methods research) has become popular. It is important, therefore, that one be able to

S.M. Lynch, *Using Statistics in Social Research: A Concise Approach*,
DOI 10.1007/978-1-4614-8573-5_3, © Springer Science+Business Media New York 2013

understand both qualitative and quantitative methods if one is to be up-to-date and well-read in one's areas of interest.

In general, the specific research question that one develops within a topic area determines the type of data and method needed to answer it. It would be impossible, for example, to investigate racial differences in obesity rates across age using a qualitative approach. Similarly, it would be impossible to fully capture the totality of the experience of becoming obese with quantitative data. Qualitative and quantitative research are often complementary: in some areas of study, it may be difficult to develop hypotheses to test quantitatively before conducting exploratory, qualitative research to understand the topic. In short, neither qualitative nor quantitative research is naturally superior to the other. Furthermore, when done well, *neither is easier to do than the other.*

3.1 Qualitative Data Acquisition

There are three broad types of qualitative research: ethnography, interviewing, and comparative-historical research, although the three overlap with each other. Ethnographers study people within their social worlds and attempt to understand and explain behavior based on the confines or rules of their world. Good ethnographic research is usually "thickly-descriptive," involves a small number of cases, and requires considerable immersion in the social world of interest. Ethnographic data is typically collected via participant-observation methods, with the degree to which the researcher becomes involved in the social world determining whether the data collection leans more toward the "participant" or "observation" end of the spectrum. On the "participation" end of the spectrum, ethnographers may fully embed themselves in the social world they are studying, like by joining a gang and engaging in their activities or living in a homeless shelter or on the street. "Observation" might be based on living in a community without taking on any formal participatory roles, but it could also be based on systematic examinations of public behavior wherever one goes. Interviewing is usually based on questions asked in face-to-face meetings. These interviews are somewhat different from those conducted by quantitative researchers, because the questions tend to be more open-ended, and key data are often quotes taken from transcriptions of the interviews.

Comparative-historical researchers attempt to understand large-scale socio-historical processes. In doing so, they often compare societies' experiences in order to reveal patterns. For example, a comparative-historical scholar may be interested in understanding why some countries more easily adopt democratic governments than others, or s/he may be interested in understanding why one country's political system evolved as it did. In order to answer such a question, a researcher may focus on historical documents or on interviewing key political figures. Some may also collect quantitative data and employ quantitative methods. Although comparative historical studies tend to be based on small numbers of cases, they tend to share more in common with statistical ways of thinking than other forms of qualitative data (Duneier 2012). This is because they tend to be interested in making claims about cause and effect. In recent years, some historical sociologists have moved in the direction of studying such topics as war, nationalism and state formation using quantitative data sets that cover the entire world over long periods of time (e.g., Wimmer 2013).

3.1.1 The "Unit of Analysis"

A key distinction between ethnographic and comparative-historical research is the *unit of analysis*. The unit of analysis is the level at which observations are made. The unit of analysis in ethnography is almost always the individual,[1] whereas the unit of analysis in comparative-historical work is always larger than the individual, e.g., states, countries, regions. As a general rule, when the unit of analysis is the individual, we call the research *micro* level research; when the unit of analysis is larger than the individual, we call the research *macro* level research. Some scholars have distinguished a third level—the "meso" level—which is intended to represent small groups or organizations. However, for our purposes, we will limit our discussion to micro and macro units.

Quantitative data may be collected at either level. Some examples of micro level quantitative data include measurements of individuals' ages, heights, weights, earnings, etc. Some examples of macro level data include state abortion rates, the proportion of the population in a state that is poor, the US unemployment rate at different points in time, etc. Notice that, while these characteristics may be aggregated from information on individuals, they are not characteristics of individuals, but of larger units, like counties, states, countries, etc.

The research question almost always determines the unit of analysis that should be investigated. For example, if we are examining whether the division of labor in society produces anomie, data must be collected at the macro level. The division of labor is a macro level phenomenon: it is not a characteristic of an individual. On the other hand, if we are examining whether poverty leads to psychological depression, the unit of analysis must be the individual.

Failing to use the appropriate unit of analysis when addressing a research question potentializes one of two fallacies: the ecological fallacy and the individualistic fallacy. The ecological fallacy is the misattribution of a relationship observed at the macro level to the micro level. The individualistic fallacy is the opposite: the misattribution of a micro level relationship to the macro level. To make these ideas concrete, consider the relationship between socioeconomic status and heart disease. Macro level studies find that richer countries have higher rates of heart disease mortality than poorer countries. Micro level studies, on the other hand, find that

[1] Often, the focus may be on the situation or conditions in which the respondent exists, but the unit at which this is examined is the individual.

	Republican	Democrat	Marginal
White	60	20	80
Nonwhite	20	0	20
Marginal	80	20	100

Table 3.1. Cross-tabulation of hypothetical political party votes by race.

	Republican	Democrat	Marginal
White	80	0	80
Nonwhite	0	20	20
Marginal	80	20	100

Table 3.2. Alternative cross-tabulation of hypothetical political party votes by race.

richer persons have lower levels of heart disease mortality than poorer persons. Thus, if we were to compare rich countries' levels of heart disease mortality to poor countries' levels of heart disease mortality and then conclude that, because the relationship between wealth and heart disease mortality is positive, richer individuals are at greater risk for heart disease than poorer individuals, we would be committing an ecological fallacy. Similarly, if we were to observe, at the individual level, that wealthier people are less likely to have heart disease than poorer persons, and conclude that wealthier countries must therefore have lower levels of heart disease than poorer countries, we would be committing an individualistic fallacy.

To demonstrate, quantitatively, a type of ecological fallacy, consider the following scenario. Suppose we know that a voting district is racially divided so that 80 % is white and 20 % is nonwhite. Suppose we also know that, in a recent election, 80 % of the district voted for a Republican candidate, while 20 % voted for a Democratic candidate. We might be tempted to conclude that whites voted for Republicans while nonwhites voted for Democrats, especially if we observed that same pattern in district after district. However, we cannot conclude this. Consider the data in Table 3.1 presented in a contingency table format—also called a cross-tabulation or "cross-tab." The table shows that nonwhites in that district were far more likely to vote Republican than whites were: 100 % of nonwhites voted Republican (20 out of 20), while 75 % of whites did (60 out of 80).

Table 3.2 presents an alternative data structure that maintains the same "marginal distributions," that is, the same proportions in the margins of the cross-tab, but the cell counts within the table differ. Under this scenario, the data do, in fact, support the conclusion we may have fallaciously jumped-to. However, as we saw, the data did not have to follow this pattern.

Technically, this ecological fallacy is of a slightly different form than the original one presented above. In fact, there are at least four forms of ecological fallacies, but their root is the same: individual inference cannot be made from aggregate data.

The data above can be "reversed" to illustrate an individualistic fallacy. Suppose we had numerous voting districts, all exactly like that shown in Table 3.1. We would observe that blacks were much more likely to vote for Republicans than whites were. We might then conclude that the proportion of voters that is black in a district is positively related to the proportion voting Republican. However, this clearly isn't the case, because all districts are 80 % white but voted Republican.

3.2 Quantitative Data Collection

Quantitative data are collected in a variety of ways. If the researcher collects the data him/herself, the data is considered "original;" if the researcher uses extant data, the data are considered "secondary." Whether the data are original or secondary to the researcher, such data are usually collected via one of three ways:

1. Face-to-face interview
2. Mail survey
3. Telephone survey

All three modes of data collection involve a written questionnaire (called an "instrument") with questions that usually have predetermined response categories. Each mode has its relative advantages and limitations, as Table 3.3 summarizes (1 is best; 3 is worst). As the table shows, mail surveys are the cheapest and the least labor intensive (one can implement a mail survey by oneself), but they are also slow and tend to produce low response rates (more on the importance of response rates later). In contrast, in-person interviews are costly, slow, and labor intensive. However, they tend to obtain the highest response rates and perhaps the most detailed data. The key advantage of the phone survey is its speed. Almost all contemporary political polls are done using phone surveys, which is why news organizations tend to be able to report opinions about Presidential speeches, debates, etc. almost immediately after they are aired.

Criterion	Mail	Phone	In-Person
Overall financial cost	1	2	3
Labor Intensity	1	2	3
Speed of data acquisition	2	1	3
Detail of information	2	3	1
Response Rate	3	2	1
Reliability and validity of responses	?	?	?

Table 3.3. Rankings of different modes of data collection along several dimensions (1 = best; 3 = worst).

In terms of reliability and validity of response—these terms to be defined more precisely later—it is difficult to determine which approach is best. For some sensitive information, the in-person interview is better than a mailed questionnaire, because a respondent may be hesitant to put his/her views in writing. On the other hand, a mailed questionnaire may be better for eliciting valid responses to some types of questions, because a respondent may be afraid to vocalize an opinion if he feels the interviewer may judge him poorly for it.

3.2.1 Sampling

Once a mode of data collection has been chosen, a researcher must decide to whom to give the questionnaire. It is generally infeasible to collect data on an entire population, and so researchers usually select a sample of individuals to interview. The importance of selecting an appropriate sample cannot be overstated. The goal of a survey is generally to provide information about a large population. Statistical theory proves that, if a sample is appropriately selected, it is possible to make very precise characterizations of the population with only a handful of sample members. However, an improperly selected sample will not provide us an accurate representation of the population; at its worst, an improperly selected sample may provide us an extremely misleading picture of the population. Furthermore, statistical theory and methods that are used to justify making inference and testing hypotheses *do not apply* to inappropriately-selected samples.

In order for a sample to represent the intended population, the sample should *ideally* be a simple random sample from the population. A simple random sample is one in which every individual in the population has equal probability of being selected.

Using a random sample may seem strange: How can randomness be a desirable quality? In fact, randomness is the only way to ensure that all characteristics of a population are represented proportionally to their existence in the population. For example, suppose we had a box containing 10 marbles of 2 colors: 5 are black, and 5 are white. Suppose I don't know how many of each color there are, but I am to estimate the proportion of each color in the box from a sample of 4 marbles. There are a total of 210 unique samples that could be drawn from this collection of 10 marbles. How did I determine the number of possible samples? The number of samples is the number of ways I can draw 4 marbles from a total of 10; by definition, this is a combination (we will discuss combinations in greater depth in the subsequent chapters). The computation for the number of combinations of x items taken from a total of n items is:

$$C(n, x) \equiv \binom{n}{x} = \frac{n!}{x!(n-x)!}, \tag{3.1}$$

# white	# black	Number (percent) of Samples
0	4	5 (2 %)
1	3	50 (24 %)
2	2	100 (48 %)
3	1	50 (24 %)
4	0	5 (2 %)

Table 3.4. Color composition of samples of four marbles taken from a population of 10 marbles, half black and half white.

where generically $k!$ is read as "k factorial" and means:

$$k \times (k-1) \times (k-2) \times \ldots \times 3 \times 2 \times 1 \times 0!, \tag{3.2}$$

and $0! = 1$ by definition.

Here, there are $\binom{10}{4} = 210$ possible samples. Of these 210 samples, other combinatorial calculations can tell us how many of these 210 samples consist of different numbers of white and black marbles. There are $\binom{5}{4} \times \binom{5}{0} = 5$ ways to draw a sample of 4 black marbles out of the 5 black marbles and no white marbles, $\binom{5}{3} \times \binom{5}{1} = 50$ ways to draw 3 black marbles and 1 white marble, $\binom{5}{2} \times \binom{5}{2} = 100$ ways to draw 2 black and 2 white marbles, $\binom{5}{1} \times \binom{5}{3} = 50$ ways to draw 1 black and 3 white marbles, and $\binom{5}{0} \times \binom{5}{4} = 5$ way to draw 4 white marbles.[2] If we were to randomly select 4 marbles—meaning each one of these 210 samples is equally likely to occur—there would be a probability of $100/210 = .48$ that we would obtain a sample in which the sample distribution was the same as the population distribution (i.e., a 50/50 % split). There would be an additional probability of $(50+50)/210 = .48$ that we would obtain a sample in which the sample distribution would be off by one marble in predicting the population distribution. Thus, with a random sample, there would be a very high probability that we would be able to make a close "guess" concerning the population distribution's composition. Indeed, there is only a probability of $(5+5)/210 = .04$ that we would conclude that there are either no white or no black marbles in the population. Table 3.4 summarizes the proportion of samples with composed of different numbers of black and white marbles.

We will discuss the importance of random sampling further when we begin to discuss statistics, but this example should help provide some intuition regarding why a random sample generally provides a better representation of a population than a nonrandom sample. With a nonrandom sample, there is no way to know

[2]See Exercise 5.10 and the solution; these computations constitute the basis of the hypergeometric distribution.

how much its characteristics deviate from those in the general population; with a random sample, statistical theory tells us how much the sample may vary from the population.

Most major surveys do not involve simple random sampling, but rather some variant of random sampling, like stratified sampling or cluster sampling. In stratified sampling, the population is split into two or more groups, and random sampling is conducted within each group. This approach to sampling is often used when the research is geared to examining group differences in some quantity, and the researcher needs to ensure an adequately large sample from all groups. For example, suppose I were interested in comparing Native Americans in the US to persons of all other races. I could divide the population into these two groups and then randomly select m persons from the Native American group and n persons from the "all other races" group. In this particular case, stratified sampling would be a better strategy than simple random sampling, because Native Americans constitute a tiny percentage of the population. A simple random sample, therefore, would probably not yield enough Native Americans (if any) to make the comparisons in which I may be interested.

In cluster sampling, (1) the population is broken into a number of "clusters," (2) a number of clusters are randomly selected, and then (3) everyone within each chosen cluster is selected into the sample. For example, the US could be broken down into neighborhoods, one could take a random sample of neighborhoods and then interview everyone in each selected neighborhood. Cluster sampling is often used when the population is large enough that it may be difficult or impossible to obtain a list of all members from which to sample.

Most major surveys today use a more complicated variant of cluster sampling called "multistage cluster sampling" or just "multistage sampling." In multistage cluster sampling, the population is broken into clusters at several levels (e.g., state, county, city, neighborhood, street). A random set of clusters is chosen (e.g., several states), then, within those clusters, a random set of clusters is chosen (e.g., several counties within each chosen state), and so on. At the lowest level, say households, a random set of households is chosen and then perhaps a random member of each selected household is chosen. Each of these deviations from simple random sampling requires some statistical adjustments to the data in order to make the final sample representative of the population. In this book, we will assume that we have a simple random sample; handling data from complex sample designs is beyond the scope of this introduction (see Scheaffer et al. 2012 and Lohr 1999 for details on sampling methods and adjustments that must be made).

In order to obtain a random sample, or some variant thereof, we typically need a listing of all the members in the population of interest (called a sample frame). Such a list may be difficult to obtain, if one exists at all. The multistage cluster sampling approach helps with this problem, because it breaks the population into clusters, lists for which typically do exist (like lists of states, counties within states, etc.). By the time the selection process gets to the neighborhood level, it may be quite easy to list all the houses.

Type of List	Problem
Voter registration list	only voters & persons over 18
Utility company list	only utility users
Phone book	only those with listed phone numbers
Newspaper subscription list	only subscribers
Email address list	only those with email
County property tax lists	only property owners

Table 3.5. Types of sample frames and some problems with them.

A number of strategies exist for obtaining lists of members of a population, but we must be careful to understand what potential biases may be introduced by using them. Table 3.5 lists some possible sample frames and some problems with them. For example, a common frame used in social research is a voter registration list. Voter registration lists are a matter of public record, making them easy to obtain. However, a key limitation of this frame is that it only contains registered voters and thus does not provide us with a sample that is representative of the entire population.

3.2.1.1 Some Invalid Strategies for Sampling

There are a number of strategies for sampling that are commonly used but are inappropriate for collecting data and reaching valid conclusions concerning *any* research question. Some of these include:

- Selecting friends and/or family.
- Web surveys (respondent self-selects).
- Phone-in surveys (respondent must call).
- Using students at one college only (only represents one college).
- Stopping people on the street, at the mall, at a restaurant, etc.

Some of these survey methods *might* produce valid results but will be generalizable only to a restricted population. In other words, we can only make claims about the population that is represented by our sample frame. For example, selecting students at random from a single college provides a valid way to make inference to that specific college but not about college students in general. Consumer surveys that come with a product may provide a valid means to determine what type of person purchases a product, but not about consumers in general. Furthermore, in such a case it only represents those who purchase such a product *and* care to complete the survey. A sample in which webpages are selected at random—and their owners surveyed—*may* yield a valid sample, but it only represents persons with webpages.

3.2.2 Response Rates

Response rates for surveys need to be high—usually responses rates below 70 % are unacceptable; rates above 90 % are usually considered good. The key consequences of a low response rate include (1) reduced statistical power and (2) the potential for biases. Why a low response rate reduces statistical power is easy to see. If the sample consisted of 100 potential respondents, and only 50 answered the survey, we have less information about the population from which the sample came. In fact, with such a large reduction in the sample size it may be impossible to compare some groups, because there may not be any respondents in a subsample of interest. For example, Native Americans constitute just under 1 % of the US population; thus, in a sample of 50, we would expect to have 0 respondents who were Native American.

A more subtle problem with a poor response rate is that it potentializes biases, where a "bias" is a discrepancy between a sample estimate and the true population quantity (parameter) of interest. As a simple example, suppose the goal of a survey was to determine who would win a presidential election, but no Democrats answered the survey. The bias in that case would be clear: the survey would suggest a clear win for the Republican, but only because there were no Democratic respondents. This is called "nonresponse bias."

We can sometimes tell whether results will be biased by nonresponse. If a number of sample characteristics do not match the characteristics of the population the sample is intended to represent—e.g., the age, sex, and race composition of the sample does not match that of the population—we should be wary. Sometimes determining whether biases are likely is difficult. For example, if the survey's goal were to examine some attitudes about a particularly sensitive issue (e.g., abortion attitudes, attitudes about gay marriage, etc.), and the survey makes that clear at the outset, persons who hold an unpopular perspective may be unlikely to answer the survey. If the unpopular perspective were not related to characteristics that are known at the population level, then it may be impossible to know that estimates concerning the attitude of interest will be biased.

3.2.3 Instrument and Item Construction

Once a sample is selected and a format for the survey instrument is selected (mail, phone, in-person interview), the specific questions that are to be asked—and their format and arrangement—should be determined. The design of questions is important, as they constitute the operationalization of your concepts of interest and will therefore determine whether you are able to satisfactorily answer your research question. That is, at the stage of developing a research question and deriving hypotheses, the relationships between concepts should be made clear, but the concepts themselves are still abstract until you attach particular survey questions/items to them. It is at this point that you need to be careful in designing the items in the questionnaire in order to appropriately measure your concepts.

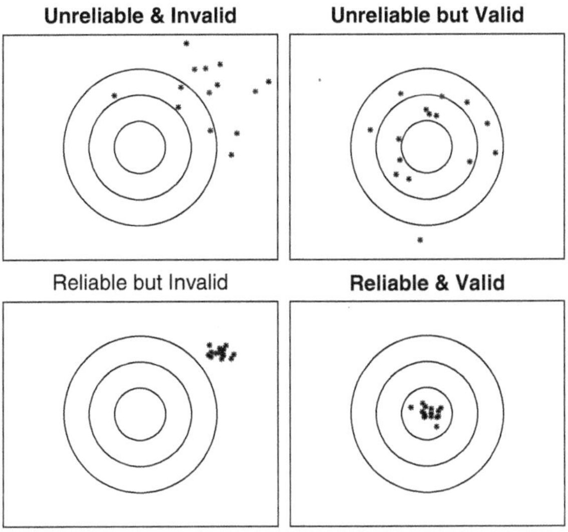

Fig. 3.1. Illustration of concepts of reliability and validity of measurement

3.2.3.1 Reliability and Validity

In the process of operationalizing your concepts via survey questions, whether you design the questions yourself or are using secondary data, you need to be sure that your measures are reliable and valid. You have much greater control over reliability and validity if you design the questions, and thus you should be careful in item construction, and ideally their location in the instrument.

Reliability can be defined as the consistency of a measure. In other words, if you ask the same question to a person again and again, or if you asked two people the same question, *all else being equal* between them, would you get the same response? If so, an item may be considered a reliable measure. Validity refers to the accuracy of a measurement, in terms of whether the question/item measures what it is intended to measure.

A classic depiction of reliability and validity is a collection of shots on a target. Figure 3.1 shows the cases graphically. If the shots are unclustered and off target, the shooter is neither reliable nor valid (upper left plot). If the shots are unclustered but centered on average over the bullseye, the shooter is valid but unreliable (upper right plot). If the shots are clustered together but away from the bullseye, the shooter is a reliable shooter (lower left plot). Finally, if the shots are clustered and centered on the bullseye, the shooter is considered reliable and valid.

Some methodologists argue that reliability is a necessary condition for validity; that is, that a measure that is unreliable cannot be valid. However, others argue, as I do here, that a measure can be a valid measure of a concept but simply not be consistent.

As an example illustrating reliability and validity, suppose we are interested in assessing individuals' health. Consider the following item:

> How would you rate your health on a scale from 1 to 100, with 1 being the worst health possible and 100 being the best?

This item seems to be a reasonable way to measure one's perceived health; however, it is not a particularly reliable measure. Why not? If I asked you this question right now, and then asked you again tomorrow, you would most likely give different answers, whether your actual health in fact varied. A number of factors, like the weather, your mood, whether you have eaten, whether you have just received news that you failed a test, etc., may affect your response, although your health hasn't actually changed. Similarly, it is quite likely that two people with exactly the same objective health will answer the item very differently. However, I would argue that, while the measure is unreliable, it is a valid measure: it seems to directly assess one's perceived health.

To make the item more reliable, we may consider reworking the possible response categories. For instance, we may change the possible responses to: Excellent, Good, Fair, or Poor. It is much more likely that, despite mood changes or other considerations, an individual will provide the same response day after day if, in fact, no real health change occurs. Similarly, it is quite likely that two individuals with the same underlying health would respond similarly. This health measure is very reliable, and studies have shown it to be a valid measure of health. In fact, this measure predicts mortality better than physician assessment and other objective health measures (Idler and Benyamini 1997)!

Now suppose we used the following as our measure of health:

> In general, how do you feel? (with Excellent, Good, Fair, or Poor as the response categories).

This item *may* be reliable, but it does not seem to be a valid measure of health. Asking respondents how they feel does not clarify whether you are referring to their current state of mental well-being, tiredness/alertness, their physical health overall, or their current general presence or lack of physical symptoms.

In general, developing reliable and valid measures is not easy, but fortunately, previous research can often guide you in developing an appropriate measure: use measures that have been validated and used before. In addition, there are at least five general rules to follow that can help in constructing good questions.

First, questions should not be double-barreled: ask ONE question at a time. Few things are more disconcerting to a respondent than being asked two questions at once—how does the respondent decide which to answer, especially if his/her responses to the two are inconsistent with one another? For example, suppose you ask "Do you favor the right of a woman to make decisions concerning her own body, or do you believe killing babies is acceptable?" Very few people, whether they are pro-choice or pro-life, would be able to answer this question. Part of the problem is that it asks two questions at once. Part of the problem is that its phrasing is vague (what does it mean for one to be able to make decisions concerning his/her own body?; what all does "killing babies" entail?); part is also that the question is leading.

Second, questions should not be leading; they should be value-neutral. The above question is a good example of an item that is leading. Certainly no one (or very few, anyway) believes that killing babies is acceptable, and so the item would not be a valid indicator of attitudes toward abortion. As another, extreme, example, suppose you ask: "Will you vote for candidate A in the next election despite the high unemployment rate during his first term (yes/no)?" It is fairly clear, given the way this question is phrased, how the interviewer would like the respondent to answer. This sort of value-laden language should not be used in an item, because it will either lead the respondent to answer in one way, or it will alienate him/her. Instead, choose neutral language that does not indicate your own predispositions.

Third, questions should not be posed as double or triple negatives. Suppose I ask: "Wouldn't you hate not having the ability to choose your own course schedule (yes or no)?" Without thinking about this item for some time, it would be difficult to decide how to answer this question, because it is posed as a either a double or triple negative: Wouldn't, hate, not. Part of the difficulty stems from the contraction at the beginning. Technically, a contraction at the beginning of a question like this implies "Would you not ...?" However, in common English, we generally ignore this technicality. For example, if I ask "Don't you like ice cream?," most people will respond "yes," meaning they do like ice cream. However, technically, the question asks "Do you not like ice cream," and a "yes" response implies you do not. Responses to double- and triple-negatively posed questions may therefore reflect differences in individual levels of pedantry, rather than true differences in preferences.

Fourth, make sure the question is understandable: don't use abbreviations in the question or use terminology with which the respondent may be unfamiliar. As an extension, don't ask respondents questions about issues that they will most likely not know about. If I ask "In the last 6 months have you experienced an M.I.?," many respondents will not know that I am asking whether they have had a heart attack. Substituting "myocardial infarction" for M.I. is not likely to help, either. In constructing questions, it is imperative that we ask the question so that the respondent understands what is actually being asked. As a corollary, we should not ask questions that address issues with which respondents most likely will not be familiar. For example, if I asked "Do you think Bayesian statistics is preferable to frequentist statistics?," or "Do you believe impossibility theorems are valuable," I will receive blank stares from most respondents.

Finally, ask what you intend to ask. This advice seems obvious, but it is a common source of trouble when attempting to operationalize a concept. Ultimately, this issue may have little to do with the respondent's ability to answer a question, but more to do with whether you are constructing a valid measure for the concept you are intending to measure. For example, if you are interested in the amount of money a respondent *earned* last year from work, be sure to ask the question using this terminology. Do not ask "How much money did you make last year?," because a respondent is likely to include capital gains, interest, pension distributions and other sources of income that are unearned. Or, perhaps, how much counterfeit tender s/he manufactured.

Asking a question in an appropriate fashion is only half the battle. The way in which the response categories are constructed may have just as much impact on the reliability and validity of the item as the way the question itself is asked. There are at least six rules to keep in mind when constructing answer categories.

First, response categories should be mutually-exclusive. This rule means that response categories should be non-overlapping. A respondent should not have to choose between two categories because they both contain the correct response. For example, if I ask about earnings and use ($0–$10,000, $10,000–$20,000, $20,000–$30,000, ...) as my response categories, in which category does a person who makes $10,000 (or $20,000, or $30,000, etc.) fall? Although this rule seems obvious, especially when the outcome categories are numeric, violating this rule is commonplace. For example, the income categories in the IRS instruction book for Form 1040—the individual tax form used by virtually everyone in the US to determine annual personal income tax—has overlapping income categories.

Violations of this rule when the outcome categories are non-numeric are often more difficult to spot, but they occur often, as well. Consider, for example, the following question:

> Which one of the following would you prefer to buy at your local supermarket, if all three are available?: (A) frozen corn, (B) canned corn, or (C) fresh corn.

While it is obvious that the question seeks to determine which type of corn one prefers, the categories are not truly mutually exclusive, because one may use each type of corn, basing the decision on which to buy on its use. If my recipe calls for grilled corn, I am most likely going to buy fresh corn, because corn kernels do not fit on the grill. But, if my recipe is for a soup, I am most likely not going to buy fresh corn, because frozen or canned corn is easier to use for that purpose.

Second, the set of response categories should be exhaustive. This rule means that all possible responses should be included as choices. Although this may seem obvious, it is one of the most common problems evident in surveys. Always make sure that all the bases are covered. If you ask a question about religious affiliation, for example, be sure to include "other" and "none" as possible responses. If you ask about satisfaction with one's phonograph, be sure to include "I don't own a phonograph" as a possible response (also recall rule four above: how many know what a phonograph is these days?). More subtly, make sure that, if there is a qualitative distinction between different types of "zeros"—or similar response categories—that your response categories are capable of detecting this. For example, if I asked "When was the last time you went to church?," and my possible response categories were "more than a year ago, less than a year ago but more than a month ago, less than a month ago but more than a week ago, last week, yesterday," how would a person who has never been to church respond? Based on the response set, s/he would have to pick "more than a year ago," but there may be a distinction to be made between individuals who have never been to church and those who have, but have not been to church in more than a year. Always be aware that, if you do not provide an exhaustive set of options, the respondent may pick a category that most closely resembles the appropriate response. However, this may not be a valid response the way you intended.

Third, response categories should be meaningful. This rule simply means that the categories should be constructed so that they make relevant distinctions between individuals. For example, an income item with cutpoints every $419 is not very meaningful, and is likely to be very unreliable. Virtually no one knows what his/her income is in a given year to the dollar.

Fourth, response categories should generally not allow neutral answers, especially with difficult attitudinal items. Many argue that a "don't know" or some other neutral response category should not be an option in attitudinal questions. People generally lean one way or another when it comes to an attitudinal item, and so a question should force the respondent to decide whether s/he is more on the negative or positive side of the response set. Having a middle category may make it too easy for the respondent to not choose. This advice is debatable, because some may argue that an individual may be genuinely undecided about a particular attitudinal item and forcing a decision makes the result unreliable.

Fifth, there should not be so many response categories that a clear distinction does not exist between possible responses. The most useful data from a statistical standpoint are interval or ratio level measures. As we will discuss, items with this level of measurement facilitate a much broader set of statistical analyses than items at the nominal or ordinal level. However, there are many cases in which an item with response categories at the interval level will not be reliable (or even meaningful). For example, asking how happy an individual is with life, with "very very happy, very happy, happy, unhappy, very unhappy, very very unhappy" as the possible responses, is likely to produce considerable meaningless variance at the ends of the distribution. How does one distinguish between being very very unhappy and being only very unhappy, for instance?

Finally, response categories should produce variation. At the other extreme from providing too many response categories is not providing enough, or at least not producing response categories that create some meaningful variation. For example, if I ask a question concerning the amount of income an individual earned last year, providing (1) $0–$1,000,000 and (2) more than $1,000,000 as my response categories is not likely to produce any variation in most samples. Indeed, a "variable" is something that varies; an item that has no variation is not a variable and is useless for statistical analysis.

Overall, these rules should be followed whenever one collects his/her own data using a survey. However, these rules should also be remembered when one is considering using secondary data. The fact that survey data was collected by someone else, possibly even a reputable organization, does not mean that the items were constructed well. It may also be the case that the items were well constructed for the original purpose in which they were collected, but they may not be well constructed for your purpose. Reconsider the item above asking about one's preferred type of corn. If the survey were specifically geared toward asking about soup recipes, in particular, determining whether people prefer fresh vegetables for soup-making, then the item may be acceptable as is for that purpose. However, if the survey data were to be used by another researcher to determine individuals' general preferences for fresh vs. preserved vegetables, the item would not necessarily be reliable or valid.

3.2.3.2 Levels of Measurement

The response categories determine the *level of measurement* of the items, and so, as we will be discussing throughout the remainder of the book, they determine the type of statistical analyses that can be performed with the data. There are four basic levels of measurement: nominal, ordinal, interval, and ratio.

Nominal level response categories are unordered and unrankable. For instance, sex (male = 1; female = 2), race (white = 1; nonwhite = 2), religious affiliation (Protestant = 1, Catholic = 2, Jewish = 3, other = 4, none = 5), are all nominal level measures. The response categories cannot be ranked, and numerical values assigned to them are meaningless. They are sometimes called "qualitative" variables, because they refer to qualitative, and not quantitative, differences between people.

Ordinal level response categories are ordered and therefore rankable, but the distance between the categories is not uniform. For example, the difference between "excellent" and "good" health is not equivalent to the difference between "fair" and "poor" health, despite the fact that these four responses are ordered. Most attitudinal items use ordinal response categories, and they are often called "Likert scale" items when they have ordered categories that include words like "very" or "somewhat" to differentiate the degree of agreement or disagreement with a statement.

Interval level variables have ordered responses with equal distances between response categories. For example, temperature is an interval level measure: temperatures are ordered, and there is the same difference between 30° and 32° as there is between 60° and 62°. Technically, they both refer to the same number of calories required to move a fixed volume of water from one point to the other on the temperature scale.

Ratio level measures are interval level measures that have a true 0, and thus ratios of responses are meaningful. For example, age has a true 0. Thus, we can say that a person who is 40 is twice as old as a person who is 20. Compare this to temperature: we cannot say that 50° is twice as hot as 25°.

Although we may not be able to make every item in a questionnaire an interval level measure, we should at least be aware of the potential consequences of having nothing but nominal level measures: our ability to analyze the data—and thus the conclusions we can reach—will be more limited than they would be if we used ordinal or interval level measures. As another suggestion, consider measuring similar items at similar levels of measurement. For example, if you expect to ask a number of questions about respondents' levels of happiness and perhaps combine these items after the fact to produce a scale, they should all be measured at the same level (and ideally, with the same response categories) to facilitate their direct combination.

3.2.3.3 Guidelines for Question Placement and General Survey Strategy to Increase Response

Finally, there are some basic guidelines for making a survey more desirable, from a respondent's perspective, and hence boosting response. Additionally, there are a couple of guidelines for improving the validity and especially reliability of items, based on question placement.

First, place all demographic (and other boring) questions at the end of the survey. Demographic questions (like age, sex, race, etc.) are generally boring to respondents. Ask more interesting questions at the beginning of the survey to draw the respondent in. Then, once they have invested the time to answer these more interesting questions, they will be more likely to finish the survey.

Second, be sure to guarantee confidentiality of response, especially if some items address sensitive topics. Many respondents will not answer sensitive questions without such a guarantee, and all Institutional Review Boards (IRBs; these university committees must approve any research project before it is undertaken) will require this guarantee anyway. Note that guaranteeing confidentiality is *not* the same as guaranteeing anonymity. A guarantee of anonymity means that no one can link a respondent's name to their responses. A guarantee of confidentiality, in contrast, means that the linkage between a respondent's name and responses will exist, but only select survey administrators will have access to it. Most surveys only involve guarantees of confidentiality.

Third, be sure to discuss generally how important it is for the respondent to answer all questions in the survey, but make sure the respondent knows that participation is voluntary. If the respondent does not understand why s/he is being given the survey, s/he will be less likely to respond, especially if the questions contain sensitive information. Additionally, one requirement of research is that participation is voluntary—you cannot force individuals to participate. Explicitly saying this not only fulfills your obligation as a legitimate researcher, but it also increases the probability of response. That is, individuals are more likely to actually participate if they feel they are doing you a favor rather than being forced to respond.

Fourth, ask general questions about a topic before asking specific questions. For example, don't ask 100 questions about health conditions before asking about general health. Asking the specific questions first will make the respondent think too much about how to answer the general question and will thus make the response unreliable (and potentially invalid).

3.3 Conclusions

In this chapter, we discussed the process of survey construction and sampling, two key components to quantitative data collection in social science research. The goal of the chapter was not to provide a thorough depiction of these processes, but simply to present an overview as a prelude to the main focus of the book: the application of statistics to quantitative data collected via surveys. If you intend to collect your own data at some point, you will want to read much more detailed texts on survey methodology. However, there is nothing wrong with using secondary data. In most contemporary quantitative social science research, researchers use secondary data collected for broader purposes than for the answering of a single research question.

For example, an important social science survey is the GSS, which was mentioned in the first chapter. The GSS is a face-to-face interview conducted every 2 years (beginning in 1972) with a random sample of about 2,000 US residents in each year. The questionnaire used in the survey consists of literally hundreds of demographic, attitudinal, and other questions (some of which are the same year to year so that trends can be examined), and the data have been used in literally hundreds of published studies in many disciplines addressing many different research questions. There are literally hundreds of existing surveys like the GSS, and it is generally easier to spend some time looking for one that may suit your needs than it is to go through the process of designing and implementing your own survey.

3.4 Items for Review

Be familiar with the following concepts, terms, and items discussed in the chapter:

- Qualitative vs. quantitative data
- Unit of analysis
- Micro vs. macro level research
- Ecological fallacy
- Individualistic fallacy
- Original vs. secondary data
- Simple random sample
- Stratified sampling
- Cluster sampling
- Multistage cluster sampling
- Nonresponse bias
- Generalizability
- Reliability and validity
- Levels of measurement (nominal/ordinal/interval/ratio)
- Rules for question and answer category construction
- Confidentiality vs. anonymity
- Rules for item placement and introducing the survey to the respondent
- IRB

3.5 Homework

Answer the following questions:

1. What is wrong with the following survey question:

> How old are you? (circle your response)
>
> Under 22   22–30   30–45   45–65   65+

2. At what level of measurement is the preceding question measured?
3. At what level of measurement would be a baseball batting average?
4. What is the unit of analysis in the GSS survey described in the conclusion of the chapter?
5. What is wrong (if anything) with the following argument: Individuals with the highest levels of education have the best health. Therefore, countries with the highest levels of average educational attainment must have the best average health levels.
6. What is wrong (if anything) with the following argument: Average IQ at college A is higher than average IQ at other colleges. Thus, if I chose a student at random from college A and a student at random from some other college, the student from college A would probably have the higher IQ?
7. What is wrong (if anything) with the following argument: In the US, wealthier states tend to vote Democratic in presidential elections. Therefore, rich people are more likely to be Democrats.
8. I want to gauge support for a government-run health insurance policy. With that aim in mind, what is wrong with the following question: "Would you support a health care reform bill that does not include a government-run health insurance option? (yes/no)"
9. What type of sampling would I be using if a split the US population into 4 regions (northeast, midwest, south, and west) and then randomly selected 250 persons from each region?
10. Develop a five-item questionnaire to address the following research question: Does gender discrimination exist in the US labor market? Decide what items need to be included, and consider the reliability and validity of your items.

Chapter 4
Summarizing Data with Descriptive Statistics

The purpose of acquiring data is ultimately to help us answer a research question we have proposed. Statistics is the toolkit we use to do so. As we said in the opening chapter, there are three basic goals of statistics: summarization, inference, and prediction. This chapter focuses on summarization. Summarization is the process of taking a potentially large volume of data and reducing it to a few quantities that adequately represent the data so that the data can be easily understood. For example, suppose I collected data on heights and weights of a sample of 500 people. I could simply report all of this information, but it would not be clear exactly what height and weight look like in the sample, let alone the general population, nor would it be clear what the relationship is between them. I could, however, summarize both height and weight in terms of their means and standard deviations, and I could compute the correlation between them to show how they are related. We will begin this chapter by discussing methods for summarizing univariate data, that is, data on one variable. Later in the chapter we will discuss basic methods for summarizing bivariate data, that is, data on two variables. Subsequent chapters will discuss methods for examining relationships between variables.

4.1 Summarizing Nominal Level Measures

When summarizing data, we first need to consider the level of measurement of the variable we are intending to summarize. If the variable is measured at the nominal level, there are limited summaries that can be made of it. In particular, some representation of the proportion of observations in each category of the variable is all that can be done. For example, suppose we have collected information on the race of 100 members of a sample as shown in Table 4.1. In these data, there are 84 whites, 12 blacks, and 4 members of other races. We could summarize these data by simply reporting the percentages in each racial category, or we could summarize this information graphically using a bar chart.

S.M. Lynch, *Using Statistics in Social Research: A Concise Approach*,
DOI 10.1007/978-1-4614-8573-5_4, © Springer Science+Business Media New York 2013

W	W	B	W	W	W	W	W	W	W
W	B	W	W	W	W	O	W	W	W
W	W	W	W	W	W	B	B	W	O
O	B	W	W	W	W	W	B	W	W
W	W	W	W	W	W	W	W	W	W
W	W	W	W	W	W	W	W	W	W
W	W	W	W	W	B	W	W	W	W
W	W	W	W	W	B	W	W	W	O
B	W	W	W	W	W	W	W	B	W
W	W	W	W	W	W	W	W	B	B

Table 4.1. Values of race for a (contrived) sample of $n = 100$ persons (W = white; B = black; O = other)

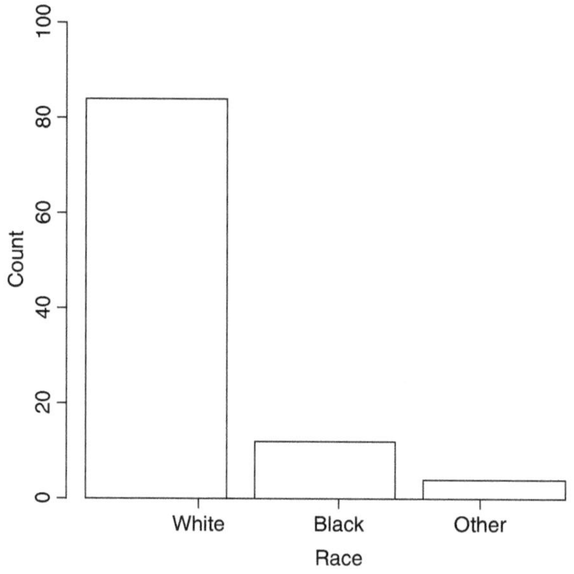

Fig. 4.1. Example of a barchart for the race data shown in Table 4.1

A bar chart provides a visual representation of data, using rectangles of differing heights to represent the proportion (or number) of cases in each category of a variable. Figure 4.1 presents a bar chart for the race data. An important part of constructing a bar chart—or any type of plot or figure, for that matter—is adequate labeling and titling. Notice that, in the figure, the three racial groups are spelled-out as "White," "Black," and "Other." Also notice that the y axis is labeled "Count." We could use "percent" instead, and that is perhaps more common in larger sample sizes. Abbreviations are not used, so the figure is immediately readable without reference to the text.

In addition to the bar chart, pie graphs can also be used to represent the proportion of observations in different categories of a nominal level variable.

Pie graphs, however, are rarely seen in scientific journals, in part because they can be misleading. Pies are always restricted to represent 100 % of a sample (or part of it), and so, in comparing the distribution of one nominal level variable (like religious affiliation) across two groups like whites and nonwhites, we would need to show two pies. These two pies would need to be of different sizes in order to reflect the different sizes of the white and nonwhite samples, and the size difference would have to be reflected in the *area* (not diameter) of the pies. This rule is often forgotten, and the result is that pie graphs often misrepresent the data they are displaying. Therefore, we will not use pie graphs in this book. Such misrepresentation is also common in figures that use symbols (like cows, or houses, etc.) that vary in size in two dimensions to represent different magnitudes. Consequently, we will not use cute symbols in the book either!

4.2 Summarizing Interval and Ratio Level Measures

When the variable we are interested in summarizing is measured at the interval or ratio level (it is numeric), there are considerably more summary measures and graphical displays that we can implement. Usually, we want to start by getting some idea of the center of the data as well as some idea about the spread of the data around its center. In statistics, measures of the center of the data are called "measures of central tendency," while measures of the spread of the data around the center are called "measures of dispersion."

Table 4.2 presents a sample of data on years of schooling for persons ages 80 and above. Our goal is to summarize these data.

4.2.1 *Measures of Central Tendency*

There are three basic statistics that are commonly used to represent the center of a set of data: the mean, the median, and the mode. The mean is commonly called the "average" outside of statistics and is computed as:

12	12	12	17	12	12	14	12	6	17
12	8	12	8	5	9	12	12	12	8
12	16	0	17	7	11	9	12	8	8
11	14	6	12	4	6	9	11	12	0
8	14	7	8	12	8	8	10	12	17
11	12	17	10	12	12	7	12	7	12
13	7	12	12	16	10	10	8	12	6
12	8	5	16	8	0	12	0	12	12
5	8	12	12	12	7	8	14	10	8
7	9	11	12	9	12	6	12	10	12

Table 4.2. Sample of $n = 100$ values of years of schooling for persons ages 80+.

0	0	0	0	4	5	5	5	6	6
6	6	6	7	7	7	7	7	7	7
8	8	8	8	8	8	8	8	8	8
8	8	8	8	8	9	9	9	9	9
10	10	10	10	10	10	11	11	11	11
11	12	12	12	12	12	12	12	12	12
12	12	12	12	12	12	12	12	12	12
12	12	12	12	12	12	12	12	12	12
12	12	12	12	12	12	12	13	14	14
14	14	16	16	16	17	17	17	17	17

Table 4.3. Sample of $n = 100$ values of years of schooling for persons ages 80+ sorted to facilitate finding median.

$$\bar{x} = \frac{\sum_{i=1}^{n} x_i}{n}, \tag{4.1}$$

where $\bar{x}$ is pronounced "x bar," the expression $\sum_{i=1}^{n}$ means the sum of x across all members of the sample (indexed by i), and n is the sample size (so all members are $x_1, x_2, x_3, \ldots, x_n$). In the years of schooling data, $\bar{x} = 10.12$.

The median is another measure of central tendency; it measures the central value in a set of data. Thus, if we have n sample members, to find the median we would sort the x from the smallest to the largest values and take the $(n+1)/2$th observation as the median if n is odd and take $[x_{n/2} + x_{(n+2)/2}]/2$ (the average of the two centermost observations) as the median if n is even. Here, n is even, and so we need the average of x_{50} and x_{51}.

The median in the years of schooling data is 11—the average of the two centermost observations (see Table 4.3). Notice that the median is not equal to the mean: In this case, the median is greater than the mean.

The third measure of the center of a set of data is the mode. The mode is simply the most frequently occurring value. From Table 4.3 it seems clear that the most frequently occurring value in the data is 12 years of education.

A stem-and-leaf plot is a good way to display the mode and also to visually represent the data to get a sense of how the data are distributed (see Fig. 4.2). In the stem-and-leaf plot, the values to the left of the vertical line are called the stems, while the values to the right are called the leaves. There are several ways to construct a stem-and-leaf plot, and the best way depends on the values in the data. Generally, the values of the stems are the leading digits in the data values; the leaves are the last digit. For example, in the last line of the plot, the 1 represents 10, and each 7 represents a value of 17 in the data. In these data there are five individuals with 17 years of education; hence, there are five "7" leaves. If, as with these data, there are only a few values for the stems (e.g., here, 0 and 1), the stems may be repeated as many times as needed, as long as we evenly divide the stems and leaves. Here, each subsequent stem and leaf row in the figure represents a one-unit increment.

```
0 | 000
0 |
0 |
0 |
0 | 4
0 | 555
0 | 66666
0 | 7777777
0 | 888888888888888
0 | 99999
1 | 000000
1 | 11111
1 | 222222222222222222222222222222222222
1 | 3
1 | 4444
1 |
1 | 666
1 | 77777
```

Fig. 4.2. Stem-and-Leaf plot of education data

Most data sets contain so many observations that it would be unreasonable to treat each leaf as representing a single individual. In such cases, each leaf may be used to represent a number of individuals, as long as this number is consistent throughout the plot.

An alternate way to represent data graphically is via a histogram. A histogram is similar to the stem-and-leaf plot, but imagine turning the stem-and-leaf plot 90 degrees so that the stems are horizontal (see Fig. 4.3). Then, imagine replacing the leaves with rectangles with heights equal to the peaks of the leaves (see Fig. 4.4).

The histogram looks very much like the bar chart presented earlier, but with a few exceptions. First, the ordering of the bars on the x axis is irrelevant in a bar chart, because there is no inherent ordering of the categories of a nominal level variable. In contrast, because a histogram shows the distribution of variables in which the categories are inherently ordered, the bars cannot be rearranged. Second, the bars in a histogram generally touch one another, while the bars in a bar chart do not. This distinction is symbolic: interval and ratio variables are generally assumed to be continuous, meaning that there is no break between adjacent categories of the variable. Put another way, all possible values between observed values are considered to exist, at least theoretically. For example, while we may only observe years of schooling in integer values, fractional values exist in theory.

The assumed continuity of values in numeric measures allows us to use an alternative approach to constructing histograms: we may replace the bars with a

```
                                    2
                                    2
                                    2
                                    2
                                    2
                                    2
                                    2
                                    2
                                    2
                                    2
                                    2
                                    2
                                    2
                                    2
                                    2
                                    2
                                    2
                                    2
                                    2
                                    2
                                    2
                        8           2
                        8           2
                        8           2
                        8           2
                        8           2
                        8           2
                        8           2
                     7  8           2
                     7  8           2
                     7  8     0     2
                  6  7  8  9  0  1  2              7
                  6  7  8  9  0  1  2     4        7
0              5  6  7  8  9  0  1  2     4     6  7
0              5  6  7  8  9  0  1  2     4     6  7
0           4  5  6  7  8  9  0  1  2  3  4     6  7
----------------------------------------------------
0  0  0  0  0  0  0  0  0  0  1  1  1  1  1  1  1  1
```

Fig. 4.3. Horizontal version of stem-and-leaf plot for education

smooth curve passing through the height of the bars from one end of the distribution to the other. Figure 4.5 shows this approach.

Returning to the measures of central tendency, with these data, the mean was 10.12, the median was 11, and the mode was 12. Generally speaking, the mean, median, and mode of a variable will not be equal. If the data follow a true normal distribution (a bell curve), then all three values *will* coincide, but this is rarely the case. Instead, sample data are typically asymmetrically distributed around the mean to some extent. The histogram above, for example, shows that the distribution of education is asymmetric, lumpy, and skewed to the left. Skewness is a property that refers to the direction in which the tail of the distribution is longer: here, the bulk of the distribution is clustered around 10–12 years of education, with a long left-hand tail that stretches back to 0. The right tail, in contrast, extends only to 17 years.

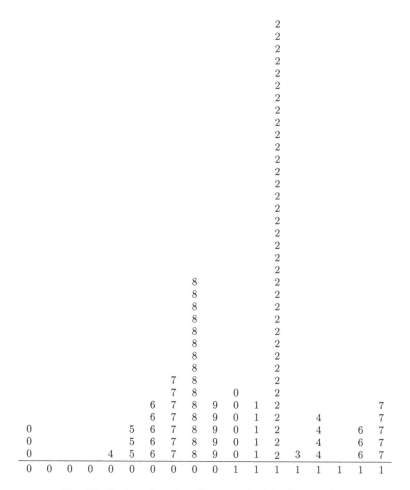

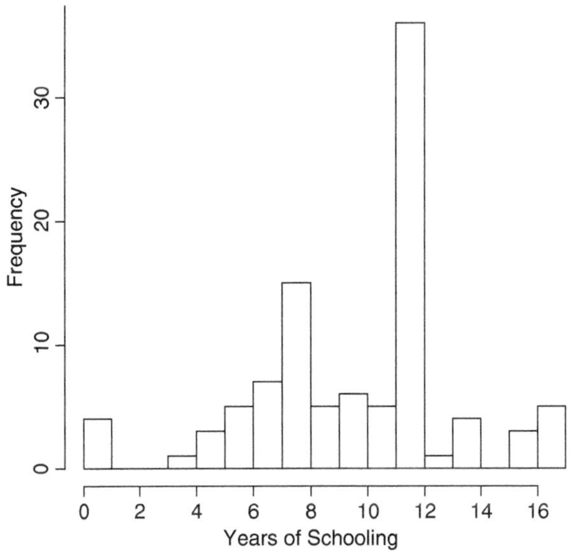

Fig. 4.4. Example of a histogram for the education data.

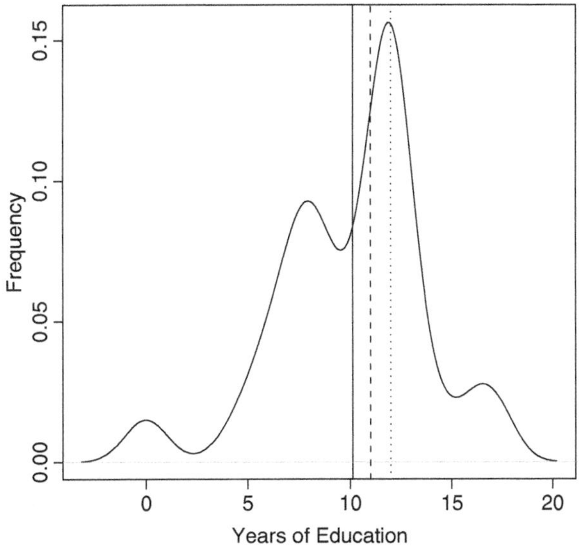

Fig. 4.5. Another approach to plotting a histogram (mean is *solid line*; median is *dotted line*; mode is *dashed line*)

Skewness is typically easy to discern graphically, but it can often be determined from the mean and median as well: the mean is *usually* pulled in the direction of the skew.[1] Thus, in distributions that are known to be highly skewed—like income—the median is usually a better measure of the center of the distribution.

The mode, in comparison to the mean and median, may not even be in the center of the distribution (although it generally is). Some distributions have more than one mode, that is, they may have multiple peaks of roughly equivalent height. In such cases, we call the distribution "multimodal." The distribution for education above is not truly multimodal, but the data do exhibit several peaks: at 0, 8, 12, and 17. The peaks at 0 and 17 are attributable to boundary issues with the response categories. There may be a number of people who had 17 *or more* years of education, but limiting the measure to 17+ years lumps everyone with 17 or more years into one category, making a peak. On the other end of the distribution, there are a number of people in older birth cohorts who received no formal education. The reason for the peak at 12 years is obvious: many people make it through high school but do not go beyond high school. The peak at 8 years is a phenomenon relevant primarily to older cohorts: among older birth cohorts, many people dropped out of school after completing grammar school (primary school) but before entering high school.

4.2.2 Measures of Dispersion

In addition to examining measures of the center of distributions of data, we usually would also like to know something about the spread of the data around the center. Some data, like education among younger birth cohorts, are fairly narrowly clustered around the center of the distribution. For example, among younger birth cohorts, most persons graduate from high school, and a large minority graduate college. Very few people do not graduate high school (although we are now seeing a reversal of this trend toward graduation), and most people do not go beyond a 4-year college degree (although there are more and more people who obtain professional degrees that require 2–3 additional years beyond the bachelor's degree). Thus, overall, education data among younger birth cohorts tend to be fairly clustered between 12 and 16 years, with very small tails of the distribution outside that range.

In comparison to education, income inequality has grown dramatically over the last 30 years, and so its distribution is quite dispersed and drastically skewed to the right. The median household income in the US is a little over $50,000, which means that half of the households in the population earn less than that. However, there are certainly households in the US that earn more than $100,000 (about 19 %; an equal

[1] I say *usually* because there are cases in which this rule does not hold. See von Hippel (2005) for more detail. Although we will not discuss skewness further here, a measure of skewness can be computed. Positive values indicate a right-skewed distribution; negative values indicate a left-skewed distribution.

spacing from \$0), and indeed, there are a few that earn well more than \$1,000,000 per year (only about 1.5 % earn more than \$200,000, however).

The purpose of measures of dispersion is to quantify the amount of spread around the center of the distribution. There are four primary measures of dispersion that are used in statistics: the range, the interquartile range, the variance, and the standard deviation. The range is the easiest to calculate: technically it is simply the difference between the maximum and minimum observed values on a variable. In the education data, for example, the minimum was 0 years and the maximum was 17. The range, therefore, is 17. However, the range is often reported as though it were an interval: 0–17.

The range is sensitive to extreme values—consider income. If one person in the population makes \$1,000,000,000 per year in income, the range is drastically larger than it would be if that person did not exist and the nearest second made only \$1,000,000 per year. Thus, a better rudimentary measure of the spread of a distribution is the interquartile range (IQR).

In order to understand the IQR, we must first define quartiles. More generally, we need to understand quantiles. Quantiles are the cutpoints that divide a (sorted) data distribution into some number of equal-sized subsets. If we are interested in breaking the data into two equal sized groups, the median is the only quantile. If we are interested in breaking the data into three equal sized groups, there are two "tertiles:" the first is the value in the data below which $1/3$ of the data fall; the second is the value below which $2/3$ of the data fall.

We can divide the data into as many quantiles as we would like, but the most commonly-used quantiles are quartiles (four subsets), quintiles (five subsets), deciles (10 subsets), and percentiles (100 subsets). As a rule, for k groups, we need $k-1$ cutpoints to divide the distribution. Thus, for quartiles, we need three cutpoints. The second cutpoint—Q_2—is the median. The first and third quartiles can be found by taking the median of each half of the data produced by dividing the data at the median.

For the education data, the median was 11. The first quartile cutpoint (Q_1) is the median of the lower half of the data, which we can find just as we found the median for the entire data set. Here, $Q_1 = 8$. The third quartile cutpoint is the median of the upper half of the data; its value is $Q_3 = 12$. The interquartile range is computed as $Q_3 - Q_1$ and is thus 4. Notice that, if we added a few more values at either end of the distribution, the IQR would most likely be unaffected. If it were affected, it would probably change only slightly, because the cutpoints are based on the ordering of the data and not so much the magnitude of the values in the data.

With the IQR defined, an additional plot can be defined that is often of use in visualizing the shape of a distribution: the boxplot (sometimes called the box-and-whisker plot). Boxplots require several pieces of information, but once this information is obtained, it is a simple plot to construct. In order to construct a boxplot, we first need the first quartile cutpoint, the median (the second quartile cutpoint), and the third quartile cutpoint. First, a box is drawn, with the first and third cutpoints being the left and right edges of the box, respectively. The median is drawn as a line within the box. After the box is drawn, the "whiskers" are added.

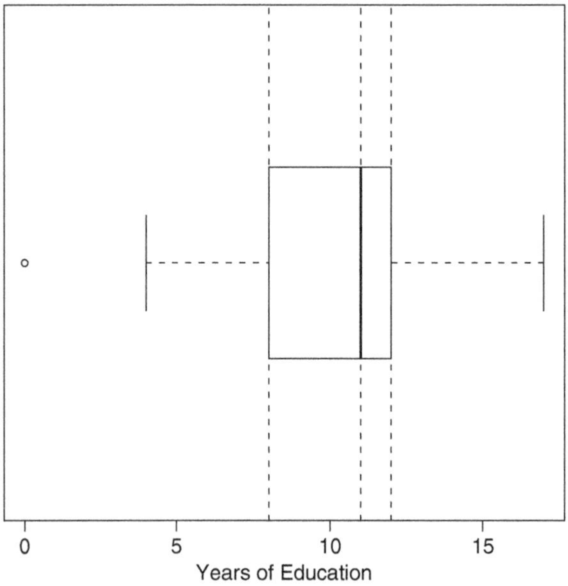

Fig. 4.6. Example of a boxplot for the education data (Q_1, Q_2, and Q_3 are highlighted with *dashed lines*).

The edge of the whiskers is usually computed as the value in the data that is within and closest to 1.5 times the IQR from the first (or third) quartile. Thus, for example, in our data, 1.5 times the IQR is 6. The first quartile cutpoint was 8, and thus, the left whisker would end at the value in the data that is closest to 2 without being beyond 2 in the direction of the tail. In our data, that number would be 4. On the other end of the distribution, the third quartile cutpoint was 12. $12+6=18$, and the largest value in the data that does not exceed 18 is 17. Thus, the values of importance in these data for constructing a boxplot are: 4,8,11,12, and 17. 4 would be the left edge of the lower whisker, 8 is the left edge of the box, 11 is the median (a line in the box), 12 is the right edge of the box, and 17 is the right edge of the upper whisker. Any value that falls outside the whisker—an "outlier"—is generally marked with a symbol like a circle or asterisk. Figure 4.6 shows the boxplot for the education data.

The last two measures of dispersion are the variance and the standard deviation. The variance is the average of the squared deviations of the data around the mean, and is denoted s^2:

$$s^2 \equiv var(x) = \frac{\sum_{i=1}^{n}(x_i - \bar{x})^2}{n-1}. \tag{4.2}$$

The denominator of this calculation is $n - 1$ rather than n for technical reasons, but the $n - 1$ ensures that our sample estimate s^2 is an unbiased estimator of

(A) Education (x)	(B) Number of Cases ($f(x)$)	(C) ($\bar{x} = 10.12$) $(x - \bar{x})^2$	(D) $(B) \times (C)$
0	4	102.41	409.66
4	1	37.45	37.45
5	3	26.21	78.64
6	5	16.97	84.87
7	7	9.73	68.14
8	15	4.49	67.42
9	5	1.25	7.27
10	6	.014	.086
11	5	.77	3.87
12	36	3.53	127.24
13	1	8.29	8.29
14	4	15.05	60.22
16	3	34.57	103.72
17	5	47.33	236.67
	$\sum = 100$		$\sum = 1,292.56$

Table 4.4. Variance calculations for education data. Dividing the sum at the bottom right by 99 $(n - 1)$ yields 13.06.

the population variance, σ^2. Notice that the numerator sums up the deviations of each observation from the sample mean, squaring this value before summing. Thus, extreme values increase the variance disproportionately compared to values that are closer to the mean. In our data, the variance is 13.06. Table 4.4 shows this computation. The table uses a convenient calculation formula, which simply involves computing the squared deviation from the mean for each unique value in the data and then multiplying by the number of times that value appears. For example, there are four values of 0 in the data; thus, we need only compute $(0 - 10.12)^2$ once and then multiply this value by four. We do not need to directly calculate this deviation four times. This computational formula can be written as:

$$s^2 = \frac{\sum_x f(x) \times (x - \bar{x})^2}{n - 1}. \tag{4.3}$$

The summation symbol, Σ simply has an index of x as shorthand to show that we are summing over all unique values of x in the data.

In addition to the frequency-based formula shown above, another formula can be derived from the original variance formula, and it involves a "trick" that will be important in various formulas later in the book. I show the derivation of this, often-called "computational method," here, which simultaneously illustrates the properties of distributing the summation symbol.

$$s^2 = \frac{\sum_{i=1}^{n}(x_i - \bar{x})^2}{n-1} \tag{4.4}$$

$$= \left(\frac{1}{n-1}\right)\left(\sum_{i=1}^{n} x_i^2 - \sum_{i=1}^{n} 2x_i\bar{x} + \sum_{i=1}^{n} \bar{x}^2\right) \tag{4.5}$$

$$= \left(\frac{1}{n-1}\right)\left(\sum_{i=1}^{n} x_i^2 - 2n\bar{x}^2 + n\bar{x}^2\right) \tag{4.6}$$

$$= \left(\frac{1}{n-1}\right)\left(\sum_{i=1}^{n} x_i^2 - n\bar{x}^2\right) \tag{4.7}$$

Equation 4.5 expands Eq. 4.4 using typical algebraic expansion ("FOIL"). The summation is simply distributed to each term. Equation 4.6 uses a couple of rules. First, multiplicative constants can be move outside of sums: $\sum_{i=1}^{n} cx_i = c\sum_{i=1}^{n} x_i$. Here, $2\bar{x}$ is a constant, and so it can be moved outside the summation, leaving $2\bar{x}\sum_{i=1}^{n} x_i$. The trick here is to recognize that, since $\bar{x} = \sum_{i=1}^{n} x_i/n$, $\sum_{i=1}^{n} x_i$ therefore is $n\bar{x}$. This leaves the center term as shown in Eq. 4.6. The last term in this equation is $n\bar{x}^2$, because we are summing $\bar{x}^2$ n times. Combining the latter two terms leaves us with Eq. 4.7. Implementing this formula is often easier than implementing the original variance formula, because we do not need to take the deviation of each observation from the mean before squaring it.

The variance is not readily interpretable, because the units are squared. A more informative measure, therefore, is the standard deviation. The standard deviation is just the square root of the variance—$s = \sqrt{s^2}$—but because it is square-rooted, its unit of measurement is the same as the original variable. This is a property that makes the standard deviation a nice measure to help summarize the spread of the data: we can say the data cluster around the mean, give or take $z \times s$ units, where z is the number of standard deviations away from the mean we want our interval to be. In these data, the standard deviation is 3.61, and so we can say that most people have 10.12 years of schooling give or take $z \times 3.61$ years of education. We will be discussing this idea in much greater depth soon, especially the meaning of, and appropriate values for, z.

4.3 Summarizing Bivariate Data

Summarizing variables one at a time is important, but we are often interested in simultaneously summarizing two variables in order to get a stronger feel for what members of a sample look like or in order to get a feel for the relationship between two variables. In terms of getting a better feel for the sample, suppose we had collected information on respondents' sexes (male and female) and races (say white and nonwhite). Univariate summaries of sex and race would be important, but it

Sex	Race		
	White	Black	Total
Male	689	118	807
	85 %	15 %	100 %
	44 %	38 %	
	37 %	6 %	43 %
Female	861	193	1,054
	82 %	18 %	100 %
	56 %	62 %	
	46 %	10 %	57 %
Total	1,550	311	1,861
	100 %	100 %	
	83 %	17 %	100 %

Table 4.5. Cross-tabulation of sex and race for the 2010 GSS.

would be more informative to have the sex-by-race breakdown of the sample. Recall from the previous chapter our discussion of the ecological fallacy: simply knowing that 49 % of the sample is male and 51 % is female, and that 80 % of the sample is white, while 20 % is non-white, does not tell us how many nonwhite males we have in our sample. As we discussed in the previous chapter, we could summarize sex and race simultaneously with a crosstab.

If the two variables we are considering are both measured at the nominal level, a crosstab is about as informative as any graphic summary we could produce. We could make a three dimensional bar chart (or an expanded two dimensional one), but the crosstab is probably the most useful summary, especially if row and column percentages are included. Table 4.5 presents a crosstab of race (whites and blacks only) and sex from the 2010 GSS with row, column, and total percentages. The format for crosstabs with percentages may vary. Sometimes only row percentages are displayed; sometimes only column percentages are displayed. In the table, I have given all three in the following order: row, column, total. It is easy to determine which is which in the table by observing which sets of percentages sum to 100 %. For example, the first set of percentages (in the first row only) are 85 and 15 %. These sum to 100 %, and so these are row percentages. They tell us the proportion of observations that are in each column, *for a given row*. Here, 85 % *of males* in the sample are white, while 15 % *of males* are black. Similarly, among females, 82 % are white, while 18 % are black.

The second set of percentages in the cells of the table sum to 100 % by column, and so they are column percentages. The table shows that, *of whites,* 44 % are male and 56 % are female. Among blacks, 38 % are male, while 62 % are female. Finally, the third set of percentages in the cells only sums to 100 % if you sum all of them: $37+6+46+10=100$ (after rounding). Thus, these are total percentages; they tell us what proportion of the overall sample falls in a given cell. For example, 6 % of the sample are black males.

How do I know whether I want row or column percentages? The answer depends on my comparison of interest. If I am interested in how genders compare in terms of their racial composition, then I would be interested in the row percentages in this table. Then I could make statements like: the ratio of whites to blacks among males is 85 to 15, but the ratio for females is 82 to 18. Thus, racial diversity is slightly higher for females than for males. This comparison of ratios might be important if, for example, I were studying the impact of imprisonment or mortality on the racial composition of the (noninstitutionalized) population. Men have much higher rates of imprisonment and mortality than women, and so the proportion of men vs. women in a sample needs to be taken into account if we are interested in how imprisonment and mortality affect racial composition. Given that the ratio for men is 85/15 (5.67), while the ratio for women is 82/18 (4.56), it seems that the mortality and imprisonment gap between nonwhite and white men may be larger than the mortality and imprisonment gap between nonwhite and white women (see Western 2009). I say "may be," because other factors, like race-by-gender differences in survey response may account for part of the difference in ratios.

If I were interested in how races compare in terms of their gender composition, then I would be interested in the column percentages here. There is a literature in sociology and economics that discusses a marriage dilemma for black women, especially for highly educated black women. The dilemma is that there are few black men available for marriage to black women, in part because of high mortality among black males, in part because of high imprisonment rates among black males, and in part because of lower average educational attainment and high employment rates among black males relative to black females (see Wilson 1987). Here, let's consider mortality and imprisonment only (since education is not in our crosstab). Evidence for the marriage dilemma can be found in the column percentages: among whites, the male-to-female ratio is 44–56 % (.79), but among blacks, the ratio is 38–62 % (.61). In other words, there are 79 white men for every 100 white women among noninstitutionalized whites, but there are only 61 black men for every 100 black women. Notice that the gender differential is somewhat severe for both races in the survey, because women are simply more likely to be respondents to the GSS interview. However, we might expect the gender difference in response to be somewhat comparable across the two races, but recall that it is possible that there may be race-by-gender differences in survey participation.

If the bivariate data one is interested in summarizing consists of one nominal level variable and one continuous variable, summarizing the data simply requires separating the data by category of the nominal level variable (called "disaggregating" the data) and computing univariate descriptive statistics. These can then be listed together in a table for visual comparison. In terms of graphic representation of such bivariate data, it is straightforward to put multiple histograms (one for each of the categories of the nominal variable) on a single plot.

If the bivariate data you are interested in summarizing consists of two numeric variables, you can either construct a scatterplot, as we will discuss later in the book, or you can categorize one of the variables and follow the strategy discussed in the previous paragraph. However, it is important to keep in mind that treating a numeric variable as nominal costs information. For example, suppose your data consisted of years of schooling and annual earnings. You could categorize education, say, as less than 12 years of schooling vs. 12 years or more, and then compute summary measures for each education group. Certainly you will find that those with 12+ years of schooling earn more than those with less schooling. But, you lose information about what type of pattern may exist across the entire distribution of schooling: It is also the case that those with 16+ years of schooling earn more than those with 12–15 years.

4.4 What About Ordinal Data?

Thus far, we have considered summarizing nominal and numeric data, but we have not discussed summarizing ordinal data. Technically, because the categories of ordinal measures can be ordered, but the spacing between adjacent categories is not necessarily consistent, measures of central tendency and dispersion are inappropriate for such data. Simply put, the numbers assigned to the categories are arbitrary. Consider the self-rated health item discussed previously in which the outcome categories were: excellent, good, fair, poor. We could assign any value to the "excellent" category, and as long as we assign increasing or decreasing values to the subsequent categories, the order of the categories would be maintained. Thus, we could assign the values 1, 2, 3, and 4 to the categories, but we could just as well assign the values 1, 2.5, 7, and 100. Both maintain the ordering of the categories, and that is all the information an ordinal measure contains. Yet these two alternate approaches to assigning values would yield extremely different values for the measures of central tendency and dispersion we have discussed. On the other hand, if we treat the ordinal data as if it were measured at the nominal level, we lose the information about the ordering of the categories.

There are a number of measures that historically have been used to describe ordinal data and especially relationships between them. However, more often than not, ordinal data are treated as numeric, and we will generally follow that practice in this book. We will assign the lowest value of an ordinal variable either 0 or 1, and assign subsequent values to categories sequentially.

4.5 The Abuse and Misuse of Statistics

In this chapter we have defined a number of measures that summarize data. Such measures can, and often are, misused either unintentionally or intentionally, and it is therefore important to be aware how they can be misused or abused. It is also important to describe some adjustments that can be made to these basic measures in order for them to be more useful for accurate presentations of data.

One of the first ways in which descriptive statistics can be misused is to use the wrong measure of central tendency when presenting data with a skewed distribution. As we discussed earlier, the mean can be misleading as a measure of central tendency when the distribution of a variable has a strong skew. For example, suppose there are 10 people in a population, with 9 of them earning $10,000 per year, and one earning $1,000,000 per year. Suppose further that, after 10 years, the person earning $1,000,000 is earning $10,000,000, while the other 9 people are still earning $10,000. One could claim that "average" income increased from $109,000 per year to $1,009,000—a factor of almost 10. Yet, the median didn't change at all. Thus, if one were interested in painting a favorable picture of a policy that generated such an income change, they may choose to report the mean. However, it would clearly be a misleading portrait. For the majority of the population, there was no improvement.

A slightly more complicated approach to presenting misleading information involving dollar amounts is to ignore inflation when reporting change over time. Inflation is a complicated topic, but everyone understands the basic idea: over time, the value of the dollar, in terms of its purchasing power, declines. When I was a child, for example, one could buy a can of soda from a machine for about $.25. In most places today, the cost is now over $1.00. For this reason, in most cases, when researchers report changes in income over time, they adjust the dollar amounts to be in stable units, like 2010 dollars. However, if one is interested in showing that incomes have increased drastically over time, s/he could simply report raw income amounts. For example, in 1970, median individual income was around $7,000 in raw dollars. By 2000, it was around $30,000 in raw dollars. Thus, it appears that income more than quadrupled. However, in fact, after adjusting for inflation, it is easy to see that incomes haven't changed at all.

Although most people would not be fooled by ignoring adjusting income amounts for inflation, many ignore inflation when it comes to other quantities in dollars, like, for example, the price of stamps. I have heard people complain about the cost of first class stamps (and I've been known to complain as well). When I was born, the price of a first class stamp was 8 cents. It is now 46 cents, a factor of almost six times as much. Yet, after adjusting for inflation, 8 cents in my year of birth translates into exactly 46 cents in today's dollars.

Aside from adjusting for inflation, dollars and other quantities that involve counts of things often need additional adjustments. In particular, *denominators* are important. The field of demography—the study of populations—is extremely concerned with adjusting raw numbers using appropriate denominators, because things that can be counted need a context in which to interpret their value (see Preston et al. 2001). For example, saying that 10,000 people per year die from some disease is meaningless unless we know how many people there are in the population. In a population like the U.S. with more than 300 million people, roughly 11,000 people die from stomach cancer each year. Yet, stomach cancer is nowhere near being a leading cause of death (it isn't even in the top 30). An appropriate adjustment to contextualize stomach cancer deaths, therefore, would be to divide by the population size. If you do so, you find that the proportion of the population that dies from stomach cancer each year is .000035—less than one 100th of 1 %. Given such a small number, a demographer might multiply it by 100,000 and call the result—3.5—the death rate per 100,000 population per year.

While few people attempt to paint stomach cancer as a major health problem, you can certainly find other conditions that produce a comparable number of deaths per year for which advocacy groups may use that number to portray it as a serious problem. The fundamental problem with using such raw numbers is that any large population will necessarily produce large numbers of deaths (or any event) from even small actual rates. Furthermore, comparing two populations of unequal sizes will always make the larger population look much worse (or much better, depending on the measure) if no adjustment is made for population size differences. Sometimes, adjustment also needs to be made to compensate not just for population size, but also age structure. For example, the crude death rate, defined simply as the number of deaths in a year divided by the number of people in the population at mid-year, is nearly twice as high in the U.S. as it is in Mexico. However, this is simply because the U.S. has a much older population: median age in the U.S. is just over 37 years, while median age in Mexico is just over 27 years.

In addition to the failure to use a denominator in presenting statistics, it is common to see advocates use an *improper* denominator to alarm people or to convince them to purchase a product that may not be that important, ultimately. One improper denominator is the clock. We could, for example, say that a person dies from stomach cancer every 48 min in the U.S. Even more startlingly, thinking about all-cause mortality, someone dies every 12 s in the U.S. from *something*. The problem with using time by itself as the denominator is that the clock never changes size, even though populations do. A population half as large as the US, but in which the proportion of deaths due to stomach cancer is exactly the same as in the U.S. (so, 5,000 deaths), will have deaths half as often. In other words, large populations will always look worse than smaller populations when time is the denominator. As a specific example, South Korea has approximately the same number of deaths from stomach cancer each year as the U.S. Thus, we can say South Korea experiences a stomach cancer death every 48 min as well. However, the population of South Korea is about 50 million—one-sixth of the U.S—and so its stomach cancer death rate is over 21 per 100,000. In short, stomach cancer is a major killer in South Korea, but not in the U.S., but using an inappropriate denominator distorts this reality.

Time is an important *part* of the denominator in some cases. For example, when measuring traffic accident fatalities, we may use miles driven *per year* as the denominator. Indeed, if one wanted to compare death rates across modes of transportation (like planes, trains, and automobiles), one would need to use miles traveled per some time unit to make the measures comparable. In demography, person-years are often the denominator for rates. Consider, for example, that, if 10 people die during the course of a year in a population that began with 100 persons, the deaths probably did not occur all at the end of the year. Thus, out of 100 persons who began the year alive, not all of them lived for a full year, and so a natural denominator for calculating the rate of death would involve person-years lived. If we assumed that, on average, those who died did so in the middle of the year, then there would be 95 person-years lived by the 100 people, and the death rate would be 10/95. Over a 1 year period, the denominator may not be much different, regardless of when the deaths occurred. However, consider how different the results might be if one were interested in some rate over a 5 or 10 year interval and all deaths occurred in the first year versus the last.

Sometimes, denominators can be constructed correctly, but results can be misleading because, in forming ratios to compare two groups (called "relative risk ratios"), the denominators drop out of the calculation. This problem is common in epidemiological work in evaluating causes of disease in which risk ratios are often constructed. For example, suppose you were told that nail biters have ten times the rate of tongue cancer deaths as non-nail biters. My agenda might be to encourage you to stop biting your nails (perhaps I'm selling a product that helps with that), and a factor of 10 seems large. However, tongue cancer has one-fifth the death rate each year as stomach cancer. Thus, if 2,000 people per year die from tongue cancer, then about 1,800 will have been nail biters, and about 180 will have not been nail biters. While that *relative* difference might seem dramatic, the *absolute risk* for both groups is small and possibly not worth considering.

This problem permeates public health promotion campaigns, as well as pharamaceutical advertising. Yet, it occurs in far more settings. Consider the issue of racial/religious profiling by airport security. Many support such profiling on the basis that "Muslims" are more likely than "Christians" to be terrorists. While that may or may not be true, what is the absolute proportion of either population that are terrorists? I would suppose it is far less than 1 % for either group. Thus, is it really reasonable to base a policy on the ratio? Similarly, keep in mind that you are *infinitely* more likely to win the lottery if you buy a lottery ticket than if you don't. Does that therefore merit your buying a ticket?

Each of the aforementioned problems essentially concerns using inappropriate measures, or inappropriately adjusted measures, of central tendency to mislead, either intentionally or otherwise. Measures of dispersion can be equally misleading, most often when they are ignored. Measures of dispersion exist to provide some sense of how much variation exists in the population; focusing exclusively on measures of central tendency can lead to misunderstanding the phenomenon of interest. For example, the average high temperature in Oklahoma is about 70 ° (F), which sounds fairly mild, but it is certainly worth knowing that Oklahoma has several weeks' worth of days over 100 ° and a few months of sub-freezing temperatures each year before deciding to move there with a limited wardrobe.

More seriously, those who have a vested interest in showing that one group is worse (or better) than another on some characteristic often fail to mention variation when reporting group differences. For example, it has been long claimed that there are racial differences in IQ, based on a difference in mean IQ between whites and blacks in the U.S. of about 5–10 points (see Dickens and Flynn 2006). Yet, this focus on the mean ignores that the variation that exists within each group produces such an overlap that the distributions of IQ are more similar than different. Given the importance of considering both central tendency and dispersion in order to evaluate whether differences between groups are meaningful, we will focus much of our attention on this topic in the remainder of the book.

We could fill up many pages discussing ways in which very basic statistics can be abused; indeed, books have been written on the topic (e.g., Hooke 1983; Huff 1993). The ease with which they can be abused is partially to blame for the claim that you can prove anything with statistics. Hopefully, this section, however, has been sufficient in showing that there are clearly wrong ways to present statistics and has encouraged you to become a more critical consumer of them.

4.6 Conclusions

In this chapter, we discussed several measures used to summarize potentially large amounts of data via only a few numbers. For nominal level data, there are relatively few possible summary measures and graphics. For numeric data, there are more measures. For this type of data, measures are divided into two groups: those that reflect the center of the distribution of data and those that reflect the spread of the data around its center. In addition, we discussed several plots that are incredibly useful for providing immediate, visual summaries of data. We then extended our discussion to summarizing data on pairs of variables simultaneously using cross-tabulations. We will illustrate the summaries repeatedly throughout the remainder of the book.

4.7 Items for Review

Be familiar with the following concepts, terms, and items discussed in the chapter. Be able to perform the calculations and produce the plots listed.

- Goals of statistics (summarization and inference)
- Measures of central tendency (mean, median, mode)
- Measures of dispersion (range, interquartile range, variance, standard deviation)
- Quantiles
- Bar chart
- Stem-and-leaf plot
- Boxplot
- Histogram
- Skewness (right/left)
- Cross-tabulation (row and column percentages)

4.8 Homework

Below is a tiny subset of the GSS data. Use these data for the following questions.

Person	Age (in years)	Sex (1 = male; 2 = female)	Education (in years)	Health (0 = poor...3 = excellent)
1	40	2	13	3
2	29	2	12	1
3	55	1	16	3
4	74	1	12	0
5	67	2	9	2
6	19	1	11	2
7	41	2	12	1
8	24	2	10	3
9	41	2	13	2
10	21	2	12	2
11	39	2	14	2
12	85	2	12	1
13	71	2	10	2
14	18	1	11	2
15	30	2	14	3
16	30	2	10	3
17	41	2	16	2
18	60	2	9	1
19	31	2	12	3
20	56	2	12	2

1. Construct a histogram for health.
2. Construct a stem and leaf plot for education.
3. Construct a boxplot for education.
4. Compute the mean, median, and mode for education.
5. Is the distribution of education symmetric or skewed? If it is skewed, in what direction is the skew?
6. Compute the mean, median, and mode for age.
7. Compute the range, IQR, variance, and s.d. for education.
8. Compute the range, IQR, variance, and s.d. for age.
9. Construct a crosstab for sex and health and interpret.
10. Split the sample into those who are above vs. below the median age. Then, compute the mean and standard deviation of health for both age groups. How does the distribution of health compare across the two age groups?

Chapter 5
Probability Theory

In the last chapter, we discussed the first goal of statistics: summarization. Summarization of data is an incredibly important aspect of statistics, but our goal in summarizing sample data is not usually to simply to report the characteristics of a sample. Instead, we are usually interested in using our sample to draw some conclusions about the population from which the sample was drawn, *and to place limits on the conclusions we can reach*. This is the role of statistical inference. For example, suppose I had a sample of 50 persons drawn at random from the population and had measured the heights and weights of all of the sample members. Suppose that the mean height was 70 in., with a standard deviation of 5 in. As we discussed in the previous chapter, if the sample is random, then our sample mean ($\bar{x}$) and standard deviation (s) may be a good guess about the population mean (μ) and standard deviation (σ). However, it is unreasonable to expect that this sample mean would be a perfect reflection of the average height in the population, because we may have a few people in our sample who were unusually tall (or short). In other words, every sample we draw from a population is likely to have slightly different means and standard deviations. The goal of statistical inference in this example would be to attempt to quantify our uncertainty about the true mean and standard deviation in the population, given the sample data that we have. Thus, we might end-up with a statement like: we are 95 % confident that the mean height in the population is 70 in., give or take an inch. In common language, this "give or take an inch" is called the "margin of error," as we will discuss in more detail in the next chapter.

Statistical inference relies on probability; in fact, inference simply inverts probabilistic reasoning. As we will discuss, probabilistic reasoning involves knowing something about a population and using that information to *deduce* characteristics of samples. Inference involves knowing something about a sample and using that information to *induce* (i.e., infer) something about the population. For example, if we know a population mean, we can use probability theory to determine the most likely values for the means of samples drawn from that population. But what if we

S.M. Lynch, *Using Statistics in Social Research: A Concise Approach*,
DOI 10.1007/978-1-4614-8573-5_5, © Springer Science+Business Media New York 2013

have a sample mean? We can "reverse" the probabilistic approach and infer what the population mean is most likely to be using statistical theory. This process forms the basis for testing hypotheses in statistics.

5.1 Probability Rules

When we refer to probability, we are generally talking about the chance of some event occurring in a trial or experiment. This definition of probability is insufficient, because it simply begs the question: isn't "chance" defined in terms of probability? One way of conceptualizing what we mean by probability is to imagine an "experiment" like flipping a (fair) coin. When we flip a coin, there are two possible, and equally likely, outcomes of the experiment (heads and tails), and the chance of obtaining a head is the ratio of the number of possible ways to obtain a successful outcome (a head) to the total number of possible outcomes. It is common knowledge that, with a fair coin, the probability of obtaining a head on a single flip is $1/2$. This conclusion is reached because (1) there are two possible (equally likely) outcomes (heads and tails), and (2) heads constitutes one of them. So, there is a one-out-of-two chance of obtaining heads on a given flip. Similarly, with a single roll of a six-sided die, there is a one-out-of-six chance of obtaining a three. It is not always the case that all outcomes are equally likely, as in these examples, but we will discuss this momentarily.

Given this basic view of probability, it is clear that probability involves (1) counting the number of ways a "success" can be obtained, (2) counting the total number of possible events that can occur in a given "trial," and (3) forming a ratio of the two. From this basic conceptualization, several terms can be defined and rules derived:

1. The collection of all of the possible events $E_1 \dots E_n$ that can occur in a given trial is called the "sample space," which is denoted as S. We denote the probability of event i occurring as: $p(E_i)$.

   Regarding coins, for example, the sample space is $S = \{\text{Heads}, \text{Tails}\}$. Some measure of the size of the sample space forms the denominator of our ratio indicating probability. Here, the size of the sample space is 2 equally likely events.
2. Probabilities are bounded between 0 and 1. An event that will definitely occur has a probability of 1, and an event that will definitely not occur has a probability of 0.

   Obviously, if a probability is a ratio of the count of ways a success can be obtained out of a given number of possible outcomes, the ratio cannot be larger than 1: there cannot be more events labeled as "successes" than there are total events possible. Also, if an event cannot occur in a trial, then the implication is that the event does not exist in the numerator. Thus the probability of such an event occurring is 0.

3. The sum of the probabilities of all the events in a sample space must be 1 ($\sum_{E_i \in S} p(E_i) = 1$).

   This rule should be intuitive: If the sample space consists of all possible events that can occur in a trial, one of the events *must* occur if we go through with the trial. So, the probability that one of the events occurs is 1.

4. The probability that two events A and B both occur—a "joint probability"—is represented as $p(A, B)$ and equals $p(A) \times p(B)$ if A and B are *independent,* that is, if the occurrence of A has no bearing on the occurrence of B. For example, if I flip two coins, whether I get heads on one coin has nothing to do with obtaining a heads on the other. Thus, the probability of obtaining two heads is the product of the probability of obtaining heads on each:

$$p(H, H) = \left(\frac{1}{2}\right)\left(\frac{1}{2}\right) = \frac{1}{4}. \tag{5.1}$$

5. When two events are not independent, $p(A, B) = p(A \mid B)p(B)$, where "$p(A \mid B)$" is the *conditional probability* of A, given that we know B has occurred.

   This rule is often rearranged algebraically to appear as:

$$p(A \mid B) = \frac{p(A, B)}{p(B)}. \tag{5.2}$$

   In this arrangement, the equation says that the probability that an event A will occur, given that we know B has occurred, is the probability that both events will occur, divided by the total probability that event B would occur. This division essentially reduces the sample space for A to the sample space that only includes B, which we already know to have occurred.

6. The probability that, of two events A and B, at least one will occur (the "union" of two events), is $p(A \cup B) = p(A) + p(B) - p(A, B)$.

The Venn diagram in Fig. 5.1 helps us understand some of these rules. For example, why do we subtract $p(A, B)$ in the last rule? We do so because it is added twice when we sum the probability of A and B. So, based on the diagram, the probability of being male is .5 (the entire lefthand circle), and the probability of being obese is .3 (the entire righthand circle). The probability of being in either circle (male *or* obese) is not .8, because that would double count the overlap region of the two circles. Instead: $p(M \text{ } or \text{ } O) = .5 + .3 - .1 = .7$.

In terms of the joint probability rule, we can rearrange the rule for non-independent events ($p(A, B) = p(A \mid B)p(B)$) as a conditional probability rule:

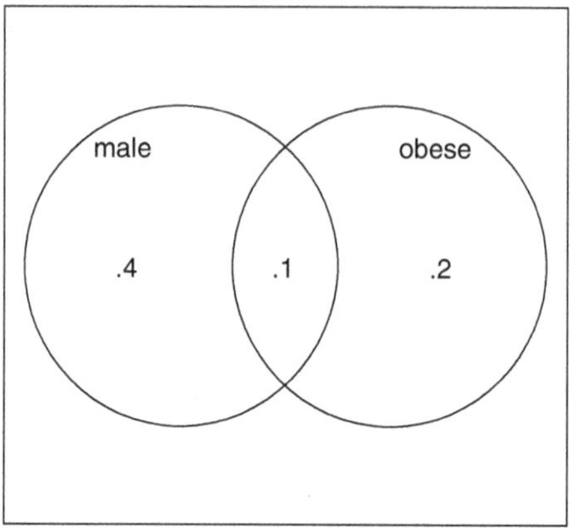

Fig. 5.1. Sample Venn diagram

$p(A \mid B) = p(A, B)/p(B)$. The figure shows why this calculation works: if we already know B has occurred/is true (say, we have been told that the person is male), then the total sample space has been reduced from the overall rectangle in the figure to the circle for being male.

As an example, consider the Venn diagram's representation of the probabilities for being male and being obese. The probability of being male is $.4 + .1 = .5$. The probability of being obese is $.2 + .1 = .3$. The probability of being an obese male (p(male, obese)) is .1 (the overlap region). The probability of being obese if a person is male (conditional on being male) is:

$$p(O|M) = \frac{p(M, O)}{p(M)} \tag{5.3}$$

$$= \frac{.1}{.5} \tag{5.4}$$

$$= .2. \tag{5.5}$$

In words, under the conditional formulation, we already know that the person is male, and so the probability that he is also obese is the ratio of the probability for being obese that is also within the male circle.

What is the probability of being a nonobese female? Given the rules above, we know that the total sample space must sum to 1. Thus, the portion of the diagram that lies outside the circles is .3 What does this area represent? It is the proportion that is neither male nor obese.

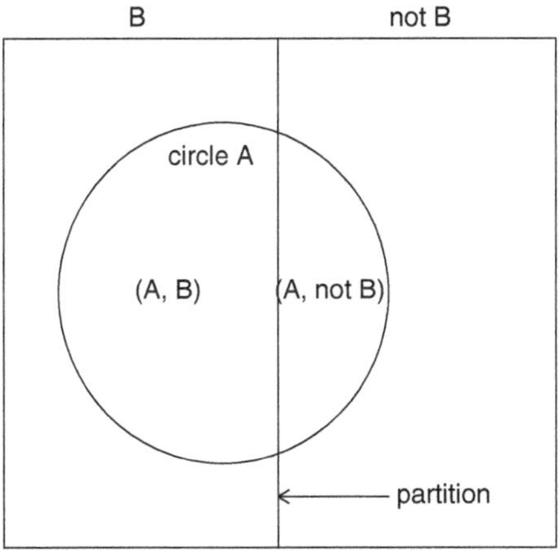

Fig. 5.2. Heuristic for the law of total probability.

5.1.1 Total Probability and Bayes' Theorem

The rules above can be used to derive more complex, important rules, like the law of total probability and Bayes' Theorem, both of which are commonly used in probability calculations. Consider Fig. 5.2, a Venn diagram with a "partition" breaking the sample space into two components—the part of the world that is B and the part that is not B ($\neg B$). If we are interested in the total probability of event A, we can use the rule for unions of events ("or"), coupled with the rule for joint probabilities of nonindependent events to compute this total probability:

$$p(A) = p(A, B) + p(A, \neg B) \tag{5.6}$$

$$= p(A|B)p(B) + p(A|\neg B)p(\neg B) \tag{5.7}$$

The first equation shows that the total probability of A is the sum of the joint probabilities of A with B and A with $\neg B$. Under the usual rule for unions, we should subtract out the joint probability of the events on both sides of the plus sign; however, here, these two events are "disjoint." That is, there is no joint probability of being in state B and $\neg B$. In words, the total probability of A is the probability of A to the left of the partition plus the probability of A on the right side of the partition.

The second equation expands the joint probability using the conditional probability rule. This equation says that the total probability of A is the probability of A occurring, given that the world is in state B, times (or weighted by) the probability the world is in state B, plus the probability of A occurring, given that the world is in state $\neg B$, times the probability the world is in state $\neg B$.

As an example of the law of total probability, suppose the weather forecast says there is a 30 % chance for rain today. I decide that, if it rains, there is an 80 % chance

I will skip class, because I don't like to walk to class in the rain. On the other hand, if it does not rain, there is a 40 % chance I will skip class, because I hate to sit in class when the weather is nice.

All things considered, am I more likely to attend class or skip it? The probability that it will rain is .3. Under that state of the world, there is a probability of .8 that I will not go to class. The probability that it will not rain is .7. Under that state of the world, there is a probability of .4 that I will not go to class. Thus, under the law of total probability:

$$p(skip) = p(skip|rain)p(rain) + p(skip|sun)p(sun) \tag{5.8}$$

$$= (.8)(.3) + (.4)(.7) \tag{5.9}$$

$$= .52 \tag{5.10}$$

All things considered, I am slightly more likely to skip class than to attend it.

The law of total probability can be extended to more than two states of the world (say n states) as follows:

$$p(A) = \sum_{i=1}^{n} p(A|B_i)p(B_i). \tag{5.11}$$

The only requirement is that $\sum B_i = 1$; that is, the probability of being in *some* state must be 1 (all the states of the world must be covered).

Bayes' Theorem uses the law of total probability and allows us to reverse conditional probabilities. What does it mean to "reverse a conditional probability?" Suppose, for example, we go to the doctor's office for a blood test for some rare disease X (occuring in 1 out of 10,000 people), and we test positive. Virtually all tests have false positive rates and false negative rates; in other words, medical tests are not infallible. So, we would like to know, in this circumstance, what our probability is for having X conditional on the positive test result. We may look online and find that the test has a 10 % false positive rate and a 10 % false negative rate. The false positive rate means that, among those who do not have the disease, 10 % will receive positive test results. The false negative rate means that, among those with the disease, 10 % will test negative. The converse of the false negative rate is that 90 % of those with the disease will test positive.

So, we know $p(\text{test} +|\text{have disease})$. We are interested in $p(\text{have disease}|\text{test} +)$. We therefore need a way to reverse the known conditional in order to obtain the probability of interest. Bayes' Theorem supplies the recipe:

$$p(B|A) = \frac{p(A|B)p(B)}{p(A)}. \tag{5.12}$$

Term	Meaning	Value
$p(A\|B)$	prob. of testing + if have disease	.9
$p(B)$	prob. of having disease	.0001
$p(A\|\neg B)$	false + (test + but no disease)	.1
$p(\neg B)$	prob. of not having disease	.9999

Table 5.1. Elements of Bayes' Theorem in disease example.

In this equation, $p(A)$ is the total probability of A, which can be found using the law of total probability:

$$p(A) = p(A|B)p(B) + p(A|\neg B)p(\neg B) \tag{5.13}$$

(or its extended version).

The theorem is easily proven by multiplying both sides of the equation by $p(A)$ and recognizing (1) the lefthand side is the joint probability $p(B, A)$, (2) the righthand side is the joint probability $p(A, B)$, and (3) these two joint probabilities are the same, only written in reverse.

Returning to the disease example, if B is "has disease", and A is "tests positive," then $\neg B$ is "does not have disease". We have all the information we need to compute $p(B|A)$ as shown in Table 5.1. Thus:

$$p(B|A) = \frac{(.90)(.0001)}{(.90)(.0001) + (.10)(.9999)} \tag{5.14}$$

$$= .0009. \tag{5.15}$$

Based on these results, even after the positive test result, the probability of having the disease (the "posterior probability") is miniscule. Why? The posterior probability is heavily influenced by two factors: (1) the marginal, or "prior" probability of having the disease, not knowing anything else (.0001), and (2) the high false positive rate. It is most likely the case that an individual does not have the disease in general, and there's a reasonably large probability (.1) under that probability that one will obtain a positive test result anyway. Put another way, we know we obtained a positive result. It is much more likely to have occured because of the false positive rate than because of the disease, given how rare the disease is.

5.2 How to Count

As noted earlier, an important part of computing probabilities is being able to count successes and the size of sample spaces. Counting seems a very basic task, but there are special ways to count in probability that often require careful thought in order to get the numerator (successes) and denominator (sample spaces) correct for computing probabilities. When trying to obtain the total number of events that may occur in a sample space, we often rely on permutation and combination calculations. We discussed combinations briefly before when discussing how to obtain a given number of colored marbles out of a box.

The key distinction between permutations and combinations is whether the order in which events occurs matters. For combinations, order does not matter; for permutations, order does matter. For example, if we have 10 persons' names in a hat and are putting together a committee of three persons drawn at random, the committee with Scott, Chris, and Brian as members would be the same committee as the one with Chris, Brian, and Scott as members. In that case, the order of selection does not matter. In contrast, suppose we were arranging these 10 people on one side of a long table. Determining how many different ways we could arrange the persons obviously involves considering the order in which individuals are placed. In that context, that's the whole point!

As we discussed earlier, the combination formula yields the number of unique sets of x items that can be drawn from a collection of n items. The calculation, again, is:

$$C(n,x) = \frac{n!}{x!(n-x)!}. \tag{5.16}$$

Let's discuss this calculation in some detail in order to differentiate how to compute combinations and permutations. The numerator of the calculation tells us how many ways n items can be arranged, ultimately into n positions. In the seating arrangement example, where order matters, determining the total number of arrangements of n people requires us to decide who will take the first seat, followed by who will take the next seat, and so on. At first, there are n persons who could take the first seat. After that person is chosen, there are $n - 1$ persons who could take the next seat. After that person is chosen, there are $n - 2$ persons who could take the next seat, and so on. Thus, there are $n!$ arrangements of the n people into the n positions at the table. This is what the numerator of the combination formula "does," and it is a very basic permutation calculation.

The denominator of the combination calculation factors out the order of (1) the persons selected out of the n persons available, and (2) the persons not selected out of the n persons.

To make this idea concrete, we must expand the example slightly. Suppose now that our table has only $x = 4$ seats. We still have $n = 10$ persons, but we now have to choose who will be selected to sit at the table and we still care about the arrangement, but only of those who are selected. In that case, we have n persons we can choose for the first seat, $n - 1$ for the second, $n - 2$ for the third, and $n - 3$ for the fourth. Thus, the number of possible ways to arrange 4 people out of 10 total is:

$$n \times (n-1) \times (n-2) \times (n-3). \tag{5.17}$$

An alternative, generic way to write this is:

$$n \times (n-1) \times (n-2) \times \ldots \times \ldots (n-x+1) = \frac{n!}{(n-x)!}. \tag{5.18}$$

In this equation, $(n-x)!$ cancels the remainder of the n persons we are not selecting. Put another way, it is factoring out of the numerator the order of the persons who are not selected to be seated at the table. Thus, another permutation calculation—one in which we are ordering x persons selected out of n persons (and discarding the rest)—denoted as $P(n, x)$ is:

$$P(n, x) = \frac{n!}{(n-x)!} \tag{5.19}$$

Let's extend this scenario one more step. Suppose now that the order in which we place the selected persons at the table does not matter; only selection vs. nonselection matters. In that case, we need to factor out the arrangement of the x persons who end up seated at the table. This involves simply reducing the problem to what remains: how many ways can we arrange the x people? Obviously, if there are $n!$ ways to arrange n people, then there are $x!$ ways to arrange x people! Thus, if we are selecting x people out of n people, and arrangement of them does not matter, we are left with the original combination formula.

All of these calculation formulas assume *sampling without replacement*. That is, once a person/object is chosen, it is no longer possible to select it again. In extremely large—or infinite—populations, combination and permutation formulas are often useless, because sampling from an infinite population is akin to sampling with replacement. However, in situations involving finite populations, these formulas are sufficient for most, if not all, problems.

To differentiate these two types of populations—finite vs. infinite—consider computing the probability of obtaining a four digit "pick 4" lottery number. In such a lottery, one pays a specified fee in order to select a four digit number, with the digits each ranging between 0 and 9. The winning lottery number is then selected at random. In a typical lottery hopper, there are four chambers, each filled with 10 balls labeled with the digits 0–9. A fan circulates the balls in each chamber, and the digit painted on the first ball that rises to the top is that position's digit in the winning number. Under that approach, there are 10 possibilities for the first digit, 10 for the second, 10 for the third, and 10 for the fourth. Given that these are independent selections, there are $10^4 = 10,000$ possible outcomes, ranging from 0000 to 9999. Thus, the probability of any given four digit number is $1/10,000$. The numbers are considered to be selected with replacement, because, even if a 1 is selected as the first digit, this selection does not affect the probability of obtaining a 1 as the second digit. It is as if there are an infinite number of 1's in the population (or as if the 1 had been replaced before the second number was picked), so the selection of a 1 on the first draw has no impact on the probability of the selection of a 1 on subsequent draws.

An alternative approach to this lottery would be that the digits are selected without replacement: that the chamber with the digits contains the digits 0–9, and that once a digit is selected, it cannot be used again. Under that scenario, there would be $P(10,4)$ possible outcomes. There are 10 possibilities for the first number (0–9), 9 for the second (whatever is left), and so on. In total, there are $10 \times 9 \times 8 \times 7 = 5{,}040$ possible outcomes instead of 10,000. Thus, if you pick a number that has nonrepeating digits, there is a 1/5,040 probability of winning in this type of lottery.

It is difficult for many people to understand that the chance, under the first scenario (sampling with replacement), of obtaining 0000 is the same as the chance of obtaining 1,234 or 5,972. This may be the result of the fact that there is a larger probability of obtaining a set of digits that do not seem to follow a pattern than the probability of obtaining a set of digits that do. In particular, clearly, the sampling-without-replacement options constitute more than half of the total possible outcomes in the sample space of sampling with replacement. In other words, of the total 10,000 possible outcomes under sampling with replacement—the largest possible set of outcomes—more than half of these outcomes (5,040) involves non-redundant digits. In contrast, there are only 10 outcomes that involve sequences of four identical digits. Thus, *as a set*, it is less likely to draw a patterned number than a non-patterned (or less obviously patterned) number, but *each* sequence itself has the exact same probability of being selected in a single lottery.

When we attempt to determine the probability for an event (or sets of events), we must decide whether counting the events in the sample space, as well as the ways successes can be obtained, requires us to consider the order in which events occur, and we must keep consistent in both numerator and denominator calculations.

Often, counting can be done either way, but the difference involves recognizing that, if order is not taken into account but should have been, that all events in the sample space may not be equally likely. For example, in a classic mistake, a famous mathematician computed the probability that one would obtain two heads on two coin flips as 1/3. His reasoning was that there are three options when flipping two coins: two heads, a head and a tail, or two tails. Yet this conclusion is based on a miscount in the denominator. When one flips two coins, there are four *equally likely* possible outcomes if order is taken into consideration: head-head, head-tail, tail-head, and tail-tail. In other words, the order matters in this situation. If one wishes to ignore order, then one has to recognize that the probability of obtaining a head and a tail is actually 2/4.

Let's consider an example involving rolling a pair of dice. Each die has from one to six dots ("pips") on its face. If we roll the pair of dice, what is the probability of obtaining a sum of seven pips between the two dice? There are two ways to answer this question. Under one approach, we would determine the complete sample space of rolls, where order matters. Following that approach, given that there are 6 possible outcomes on the first die, and 6 possible outcomes on the second die, there are 36 possible rolls. The sample space looks like:

$$S = \{(1,1),(1,2),\ldots,(1,6),(2,1),\ldots(2,6),\ldots,(6,1),\ldots,(6,6)\}. \quad (5.20)$$

The number of pairs in this set constitutes the denominator of our probability calculation. What remains is to count the number of ways a success—that is, a sum of 7—can occur from these 36 outcomes. That set looks like:

$$S_7 = \{(1,6),(2,5),(3,4),(4,3),(5,2),(6,1)\}\,. \tag{5.21}$$

There are six pairs in this set. Thus, the probability of rolling a sum of 7 on a pair of dice is $6/36 = 1/6$.

The second method of solving this problem involves recognizing that the order of the dice does not matter; for all intents and purposes, they are interchangeable. Thus, we could consider the sample space to be the possible sums that can arise. The sample space represented in this fashion would be:

$$S = \{2,3,4,5,6,7,8,9,10,11,12\}\,. \tag{5.22}$$

Under this approach, in finding the probability that we would roll a sum of 7, we must take into account the fact that these sums *are not equally likely to occur*. For example, there is only one way to roll a sum of 2: both dice must come up ones. Three are two ways to roll a three: a 1 followed by a 2 or a 2 followed by a 1. There are three ways to roll a four: a 1 followed by a 3, two twos, or a 3 followed by a one. And so on. Overall, computing the probability of rolling a sum of 7, then, involves mapping the possible events in the sample space onto a set of probabilities associated with those events and then summing up the relevant probabilities. The set of probabilities associated with the events in this sample space is:

$$p(E) = \left\{\frac{1}{36},\frac{1}{18},\frac{1}{12},\frac{1}{9},\frac{5}{36},\frac{1}{6},\frac{5}{36},\frac{1}{9},\frac{1}{12},\frac{1}{18},\frac{1}{36}\right\} \tag{5.23}$$

To some extent, even under this approach, in order to determine the probabilities of particular sums, we had to return to considering the order of rolls, but sometimes we can develop calculations (or others already have) that simplify the process. We now discuss such cases.

5.3 Probability Density/Mass Functions

When sample spaces are large, we can use algebraic functions to assign probabilities to events within the sample space. These functions must obey the rules of probability outlined above, including that all events have probabilities between 0 and 1 and that the sum of the probabilities of all events must be 1. However, aside from those key rules, the form of such functions is quite flexible. Two common such functions—called probability density functions (pdfs) if the sample space is continuous and probability mass functions if the sample space is discrete (like counting numbers;

integers)—are the binomial mass function and the normal density function. These two probability distributions are the most common ones used in statistics, and so some in-depth discussion of them is warranted.

5.3.1 Binomial Mass Function

The sample space for a single coin flip is pretty simple, consisting of only two outcomes, and so computing the probability of the two outcomes is easy. But, what if we were interested in knowing the probability of obtaining, say, 5 heads in 10 coin flips? In that case, the sample space is considerably larger. The possible outcomes (counts of heads) in 10 flips are: $\{0, 1, 2, 3, 4, 5, 6, 7, 8, 9, 10\}$. Furthermore, this is a case in which the probabilities of the outcomes are not equal. For example, there is only one possible way to obtain 0 heads in 10 tosses: Every toss must be a tail. But, consider the possible ways one can obtain 1 head. The head could occur on the first toss, the second toss, the third toss, and so on. The binomial mass function allows for straightforward computation of such probabilities. Its mass function is:

$$p(x) = \binom{n}{x} p^x (1-p)^{n-x}. \tag{5.24}$$

The random variable (the count of successes; the quantity that is random) in this function is x, while n and p are the "parameters" of the distribution. n is the number of trials, and p is the probability of success on any given trial. In mathematical shorthand, we say: $x \sim Binomial(n, p)$ ("x is distributed binomially with parameters n and p"). The sample space for x is determined by n. x must be between 0 and n, and x is restricted to nonnegative integers. Given a value for n and p, the probabilities for x can be computed. For example, if we wanted to know the probability of obtaining 3 heads on 10 flips of a fair coin, we would solve $p(x = 3) = \binom{10}{3}.5^3(1 - .5)^{10-3} \approx .12$. If we want to know the probability of obtaining a range of successes (e.g., three or fewer successes, six or more successes, etc.), we simply compute the probabilities for all relevant events and sum them.

The binomial mass function may appear somewhat complicated at first, but it involves the basic ideas of counting outlined in the previous section and the probability rule for independent events. Suppose we want to know the probability of obtaining 3 heads in a row. This computation would simply involve multiplying the probability of obtaining a head on each flip: $(.5)(.5)(.5) = .125$. Suppose instead, we want to know the probability of obtaining 2 heads and 1 tail. The probability of getting two heads on two flips is .25. The probability of getting a tail on a single flip is .5. It may seem as if we could simply multiply .25 by .5 and obtain the final answer. However, doing so ignores that the tail may come in any position in the three flips. It could be first, second, or third. In other words, there are three ways

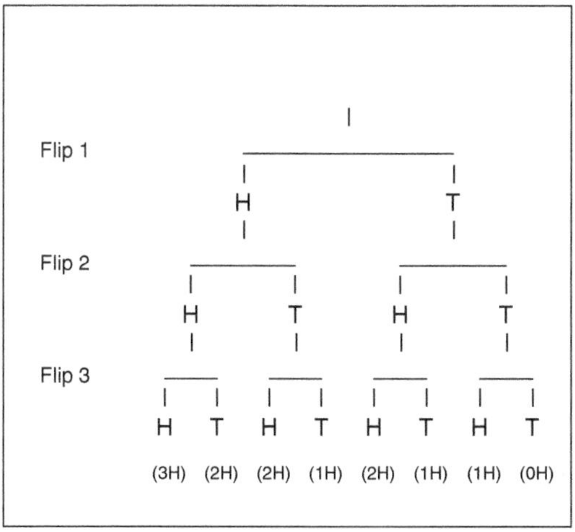

Fig. 5.3. Tree diagram showing outcomes of three coin flips

the heads and tail could be ordered. The combinatorial at the beginning of the mass function accounts for these different arrangements.

Figure 5.3 shows a tree diagram that illustrates this point. The figure shows the possible sequences of outcomes of three successive flips. The total number of heads on the three flips for each possible branch of the tree are shown in parentheses at the bottom. As the figure indicates, there is only one branch that produces three heads (similarly, there is only one branch that produces three tails—0 heads). There are three branches that produce 2 heads and three branches that produce 1 head. In this scenario, each outcome on a given trial is equally likely. Thus, each branch is equally likely, and so the probability that we will obtain two heads and a tail is $3/8$. In the event that the outcomes on a given trial are not equally likely (i.e., p is not .5), we can insert the success/failure probabilities in the branches as appropriate and multiply them along a branch to obtain the probability of a particular branch. However, this is what the righthand side of the binomial mass function does for us.

Figure 5.4 shows three examples of binomial distributions. The upper left distribution represents the probability of different counts of successes out of 10 trials when the success probability on any given trial is .5. The upper right distribution shows the probability when $p = .8$, and the bottom left distribution shows the probability when $p = .2$.

5.3.1.1 Pascal's Triangle

The combinatorial expression at the beginning of the binomial mass function is sometimes called the "binomial coefficient." For small values of n, it may be easier to use Pascal's Triangle to determine how many different ways a given count of successes can be obtained than to use the combinatorial formula. Pascal's triangle

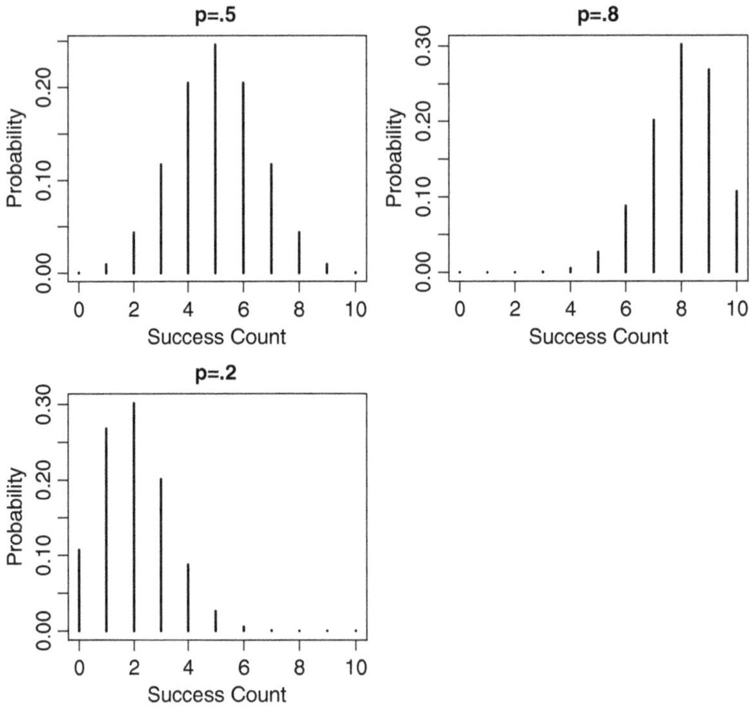

Fig. 5.4. Some binomial distributions ($n = 10$)

Trials	Ways to Obtain Sucess Counts										
0						1					
1					1		1				
2				1		2		1			
3			1		3		3		1		
4		1		4		6		4		1	
5	1		5		10		10		5		1

Fig. 5.5. Pascal's triangle. Each row shows the number of ways to obtain x successes out of n trials for different values of x, where x increases from left to right from 0 to n

is a collection of numbers with ones along the left and right edges of the triangle and sums filling in the rest of the triangle. The number at each location in a row in the triangle is simply the sum of the two numbers above it in the previous row. Figure 5.5 shows the first six rows of the triangle. To interpret the triangle, consider the last row in the figure. This row tells us how many ways there are to obtain different counts of successes (x) in a binomial distribution with an n parameter of 5 trials. Thus, there is 1 way to obtain 0 successes, 5 ways to obtain 1 success, 10 ways to obtain 2 successes, 10 ways to obtain 3 successes, 5 ways to obtain 4 successes, and 1 way to obtain 5 successes.

Combination	Equivalent	Equals
$\binom{n}{x}$	$\binom{n}{(n-x)}$	$\frac{n!}{x!(n-x)!}$
$\binom{n}{n}$	$\binom{n}{0}$	1
$\binom{n}{(n-1)}$	$\binom{n}{1}$	n

Table 5.2. Some helpful shortcuts for calculating combinations.

Pascal's triangle can be a useful shortcut to determining counts of combinations for small n. If n is large, it may take longer to draw the triangle than to simply compute the combinatorial. For large n, Table 5.2 shows some helpful shortcuts for calculating combinations.

5.3.2 Normal Density Function

Many phenomena in the natural (and social) world are distributed so that the majority cluster around some "middle" value, with more extreme cases occurring less frequently, with frequency declining rapidly with distance from the center. Consider, for example, the sample histogram for education in the previous chapter: most values of education were clustered around 12 years of schooling, with fewer persons having many more or many fewer years. Despite the slight multimodality and the skew, the distribution of schooling followed this general pattern. Many phenomena match the pattern more closely, like the distributions for height and weight in the population.

The normal distribution (also called the bell curve) represents this pattern. The normal density function is:

$$f(x) = \frac{1}{\sqrt{2\pi\sigma^2}} \exp\left\{-\frac{(x-\mu)^2}{2\sigma^2}\right\}. \tag{5.25}$$

The random variable in this function is x, while μ and σ^2 are the parameters (the mean and variance, respectively), and we say $x \sim N(\mu, \sigma^2)$ ("x is distributed normally with a mean of μ and variance of σ^2"). $\exp\{c\}$ is simply an alternative way to write e^c, with e being the base of the natural logarithms—the exponential function.

Although this density function looks complicated, it simply defines a bell-shaped curve with tails that asymptote to 0 (i.e., they diminish toward 0 the further one moves from the center, but they never reach 0). Recall from algebra that a general formula for a parabola is:

$$y = a(x-h)^2 + k. \tag{5.26}$$

The point (h, k) is the vertex of the parabola, and a determines whether the parabola is narrow or wide. The interior of the "kernel" of the normal distribution (the part inside the exponential function) is simply a parabola. The x coordinate of the vertex (technically, the "abscissa") is μ, and $-1/2\sigma^2$ is a. The negative sign flips the parabola so that it opens downward. The exponential function wrapped around this downward-facing parabola bends the tails of the parabola outward (toward $+\infty$ and $-\infty$), because the exponential function always produces a non-negative result. Consequently, the bell curve sits entirely above the x-axis. The expression in front of the exponential function determines the height of the inverted parabola. If you are familiar with the exponential function, you may recall that $e^a e^b = e^{a+b}$. Thus, the leading term $1/\sqrt{2\pi\sigma^2}$ can be logged and placed within the exponential function. From there, it is easy to see that term becomes k—the y coordinate of the vertex—in the general formula for a parabola.

Notice that the left side of this function is written as $f(x)$ rather than $p(x)$ as in the binomial distribution; the reason for this is that the normal distribution is a continuous distribution, and thus the probability of any particular value for x is 0. In a continuous distribution, the denominator of the usual probability ratio (successes over sample space) is infinitely large: there are an infinite number of real numbers between any two values in the sample space. As a consequence, we cannot determine probabilities by simply computing one application of the function. Instead, we must determine probabilities for ranges of x using integral calculus. For given values of μ and σ^2, we can compute probabilities for x falling in some desired range. The domain of x is unrestricted; x can take any real value.

The normal distribution is the most important distribution in statistics, and much of our discussion regarding statistics will focus on this distribution. Although you need not memorize the density function, it is important to know a few things about this distribution. First, the mean, median and mode of the distribution are equal (μ). Second, the distribution is perfectly symmetric around the mean, so that $p(-x < -z) = p(x > z)$. In English: the probability that x falls above some value z is the same as the probability that $-x$ falls below $-z$. Third, the width of the distribution is governed by the variance parameter σ^2. Larger values of σ^2 imply a wider and shorter distribution than smaller values of σ^2. Figure 5.6 shows three normal distributions with different means and variances.

If a variable is normally distributed in the population—that is, its histogram follows a curve like those shown in Fig. 5.6—and we know μ and σ^2, then we can determine the probability of obtaining x values in any range. In order to find probabilities, we standardize our desired range for x and look up the appropriate probabilities in a "z table" like the one in Appendix A. To standardize x we simply subtract off the mean and divide by the standard deviation:

$$z = \frac{x - \mu}{\sigma}. \tag{5.27}$$

This process gives us a value z, the distribution for which has a mean of 0 and standard deviation of 1 (called the standard normal distribution). If you consider

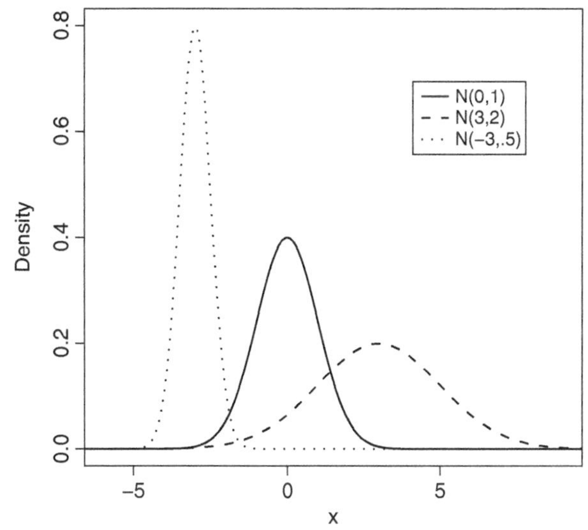

Fig. 5.6. Some normal distributions

this for a moment, z is really just a count of the number of standard deviations away from the mean that a given value of x is. Some standard probabilities from the z distribution are well-known. For example, we know (and you should memorize this) that approximately 68 % of the mass of a normal distribution falls within 1 standard deviation of the mean, 90 % falls within 1.645 standard deviations, 95 % falls within 1.96 standard deviations, and 99 % falls within 2.58 standard deviations (see Fig. 5.7). For simplicity (but not exactness), we often round the standard deviations and probabilities and use a "1, 2, 3" rule of thumb: 68 % of the mass is within 1 standard deviation, 95 % is within 2, and almost 100 % is within 3 (99.7 %). We will follow this convention throughout the remainder of the book as a matter of convenience.

For example, if I claimed that IQs were normally distributed in the population with a mean of 100 and a standard deviation of 15, then I could conclude that only about 2.5 % of the population have IQs above 130. How? If 95 % of the mass of the distribution falls within two (1.96) standard deviations of the mean, and I know that the distribution is symmetric, then there is only 2.5 % of the mass of the distribution beyond two standard deviations on either end of the distribution.

Often we are given a value of x that, when transformed to z scale, is not one of these values for which the probability is immediately known. In those cases, we simply look the z up in a z table and find the associated probability. Doing so may be a tedious process, especially given that most z tables—in order to save space—only provide probabilities for one half of the distribution. In this book, the z table provides only the probability an observation falls below a value of $-Z$.

As an example of using the z table, suppose I select a person at random from the population and want to know the probability that person has an IQ greater than 125.

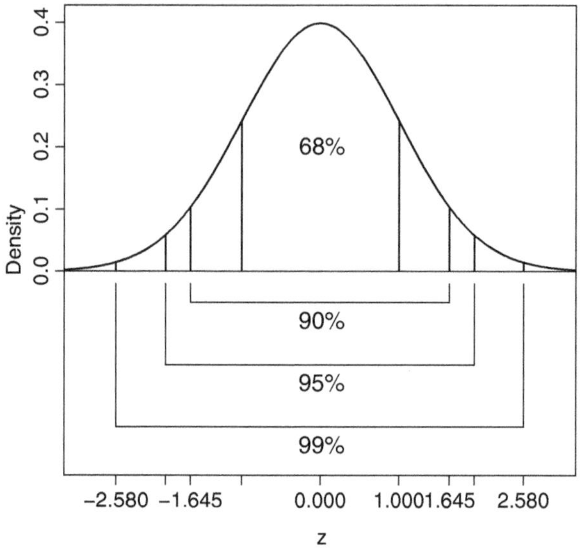

Fig. 5.7. Some common areas under the normal distribution at given numbers of standard deviations from the mean

In that case, I need to standardize the value of 125 so that I can find the probability in the z table:

$$z = \frac{x - \mu}{\sigma} \tag{5.28}$$

$$= \frac{125 - 100}{15} \tag{5.29}$$

$$= 1.67. \tag{5.30}$$

If I want to know $p(IQ > 125)$, this is equivalent to $p(z > 1.67)$. The z table in the appendix only shows negative values for z. Under the symmetry of the distribution, I know that $p(z > Z) = p(-z < -Z)$; here, $p(z > 1.67) = p(-z < -1.67)$. So, I can find $z = -1.67$ in the table. I find that $p(z < -1.67) = .047$. Again, by symmetry, if there is a probability of .047 of obtaining a person who is at least 1.67 standard deviations (z units) below the mean, then there is the same probability of obtaining a person who is 1.67 standard deviations above the mean.

Given the tedium of determining probabilities using the z distribution, I strongly recommend, whenever faced with this type of probability problem, that you take a moment to sketch a normal distribution and mark the area that you are attempting to find. Doing so is especially important when the probability an event falls in some central region is of interest. For example, suppose we would like to determine the probability that a person selected at random has an IQ between 95 and 110. In that case, we need to compute two z scores:

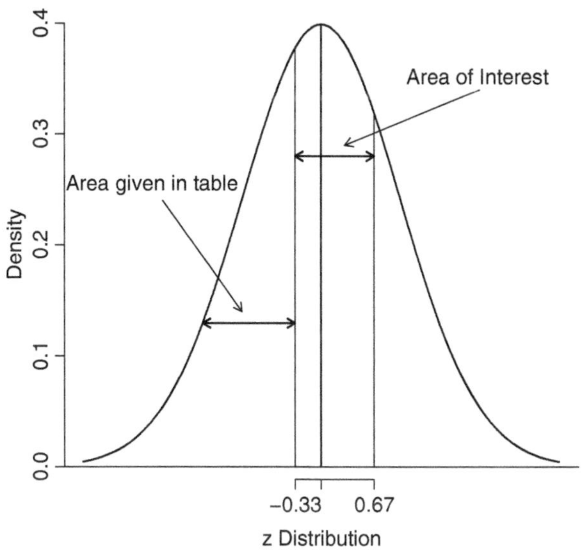

Fig. 5.8. Normal distribution area shown in z table vs. the area of interest in IQ example

$$z_1 = \frac{95 - 100}{15} \tag{5.31}$$

$$= -.33 \tag{5.32}$$

$$z_2 = \frac{110 - 100}{15} \tag{5.33}$$

$$= .67 \tag{5.34}$$

Figure 5.8 shows the region of interest and the region that the z table in the appendix provides. It is clear, from the figure, that we can find the area to the left of $-.33$ in the table, but the remaining areas must be found via using the symmetry property of the normal distribution and subtraction. From the table, the area to the left of $-.33$ is .371. Furthermore, given that $p(z > .67) = p(-z < -.67)$, the area to the right of .67 is .251. Finally, given that the total area under the normal distribution is 1, the area between $-.33$ and .67 is: $1 - (.371 + .251) = .378$. This is the probability of interest.

5.3.3 Normal Approximation to the Binomial

As we have seen in the previous two sections, computing cumulative probabilities may involve multiple computations when using the binomial distribution, but is fairly easy with the normal distribution. For example, determining the probability

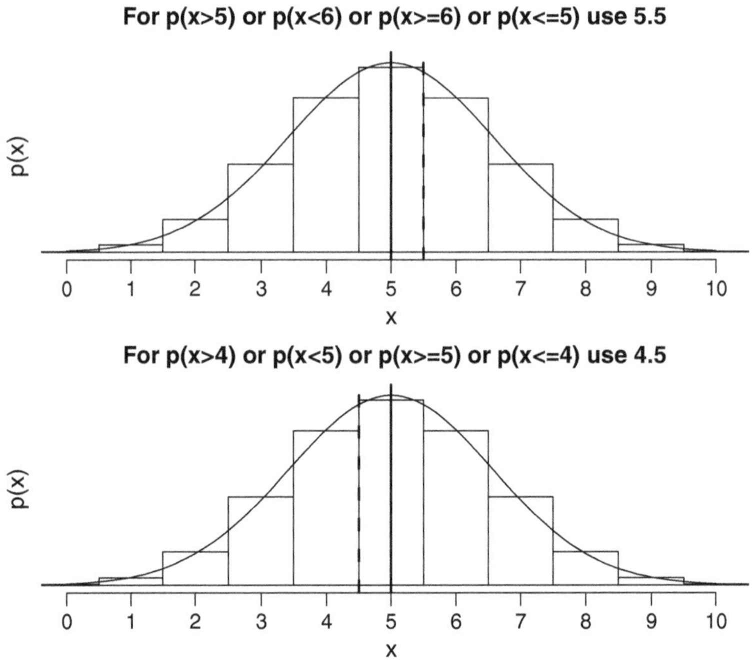

Fig. 5.9. Illustration of continuity correction factor in binomial distribution with $n = 10$ and $p = .5$

of obtaining 100 heads or fewer on 210 coin flips would involve summing the probabilities obtained from 101 applications of the binomial density function (from 0 through 100). In contrast, determining the proportion of the population that has IQs above, say, 50 would only require the calculation and evaluation of one standardized score.

In cases where the binomial distribution is unwieldy, it is possible to use a normal distribution approximation to it. If n is sufficiently "large," say 10 or so, and p is close to .5—or more, generally, if $np > 5$—then the following approximation works pretty well:

$$z = \frac{(x \pm .5) - np}{\sqrt{np(1-p)}}. \tag{5.35}$$

This equation looks similar to our previous standardized score calculation, but we have substituted np for μ and $\sqrt{np(1-p)}$ for σ. np is the expected count of successes, given a particular sample size n and success probability p. $p(1-p)$ is the variance of a proportion, while n times that is the variance of a count.

We have also added a "continuity correction" (the "$\pm .5$") to compensate for the fact that the binomial distribution is discrete, while the normal distribution is continuous. The rationale for the correction factor is displayed in Fig. 5.9. The figure shows two plots. In both plots, a normal distribution (represented via lines) is superimposed over a binomial distribution (represented via bars) with parameters $n = 10$ and $p = .5$. The normal distribution appears to match the binomial distribution quite well. However, notice that, while $p(x < 5) = 1 - p(x > 5)$ in the normal distribution because there is no probability associated with the exact value of 5, we must make a choice whether to include the probability associated with exactly 5 successes in the binomial distribution computation. The upper plot shows that, if we are interested in the probability that x is greater than 5, less than 6, greater than or equal to 6 or less than or equal to 5, then we should *add* .5 as a correction. The lower plot shows the cases in which we should subtract .5 as a correction. The logic of choosing whether to add or subtract the .5 correction factor is fairly easy to visualize, and so I recommend drawing a figure like the one shown when trying to determine the appropriate computation.

Applying the normal approximation formula is straightforward. For example, determining the probability of obtaining at least 100 heads on 210 coin flips would involve the following steps. First, realize that the probability of obtaining 100+ heads is equivalent to 1− the probability of obtaining 99 heads or fewer. Then, compute the z-score associated with $x = 99$ using the above approximation:

$$z = \frac{99.5 - (210)(.5)}{\sqrt{(210)(.5)(.5)}} = -.76 \qquad (5.36)$$

This area is approximately .224. After subtracting from 1, we obtain .776. In fact, using the binomial distribution, we would obtain .7761.

It is important to note that, if n is small, or p varies much from .5, the approximation may not work well. In particular, when p varies considerably from .5, the binomial distribution is not symmetric, *unless n is large enough to offset the asymmetry*. When n is small, even if p is very close (or equal) to .5, the normal approximation may perform poorly because of the distinction between continuous and discrete calculations, even with the continuity correction.

5.4 Conclusions

In this chapter, we discussed probability theory in considerable depth, leading up to the discussion of two important probability distributions used in statistics—the binomial and the normal. These distributions are two of the most commonly used distributions in social science research, and so we spent considerable time developing and discussing them. Importantly, we established that, if a variable follows a normal distribution, we can standardize it and evaluate probabilities of obtaining particular ranges of scores using the z table. This computation will become extremely important in subsequent chapters as we extend it for use in problems of inference.

5.5 Items for Review

- Probability and its rules
- Sample space
- Event
- Joint probability of independent events
- Joint probability of non-independent events
- Conditional probability
- Law of total probability
- Bayes' Theorem
- Probability of one *or* another event
- Venn diagram
- Combinations and permutations
- Sampling with and without replacement
- Probability density function
- Binomial distribution
- Tree diagram
- Pascal's triangle
- Normal distribution
- Standardized score
- Normal approximation to the binomial

5.6 Homework

1. What is the probability that a family with two children has two girls?
2. Now, suppose I introduce you to one of the children, and it is a girl. What is the probability that both children are girls?
3. Suppose license plates in a particular state consist of three letters followed by three numbers, and that both the letters and numbers can repeat. How many license plates can the state produce before repeating?
4. Now suppose license plates in a particular state consist of three letters followed by three numbers, but letters and numbers cannot be repeated. How many plates can the state produce without repeating?
5. Suppose a state produces licence plates that consist of four letters that can be repeated, and your name is "John." What is the probability, assuming you're the first person issued a plate, that the plate will have your name on it?
6. I have a set of three dice. One is a typical 6-sided die, one is a 4-sided die, and one is 10-sided die. First, I roll the 4-sided die. If that result is a 1, 2, or a 3, then I roll the 6-sided die. If the result of the roll of the 4-sided die is a 4, then I roll the 10-sided die instead of the 6-sided die. What is the probability, overall, that in this process I will roll a 5?
7. I rolled a 5. What is the probability that I rolled the 10-sided die?

8. If IQ is normally distributed in the population with a mean of 100 and standard deviation of 15, what is the probability of randomly selecting a person with an IQ greater than 130?
9. What is the probability of obtaining two such people in a row?
10. There are six marbles in a box; four are red and two are green. Draw a histogram showing the frequency of red marbles in samples of size $n = 3$.
11. What is the probability of obtaining 4 heads in a row, followed by a tail, on 5 flips of a fair coin?
12. What is the probability of obtaining 4 heads and 1 tail on 5 flips of a fair coin, regardless of the order of heads and tails?
13. What if the coin were weighted so that the probability of obtaining a head on any given flip were .8?
14. In the game Yahtzee, a "yahtzee" happens when you obtain the same number on each of five dice. What is the probability of getting a yahtzee on a single throw (i.e., getting 5 of the same number on one roll)?
15. Male body weight is approximately normally distributed in the population with a mean of 190 pounds and a standard deviation of 59 pounds. What proportion of males weighs between 175 and 200 pounds?
16. What is the probability of obtaining a sample of five men, all of whom are in that weight range?
17. Approximately 50 % of the population is male, and 30 % is obese. Twenty percent of males are obese. Draw a Venn diagram illustrating these proportions, as well as the proportion of the population that are non-obese females.
18. Based on the above problem, if I randomly select an individual from the population, what is the probability the person would be either a male or obese?
19. Based on the above problem, if I randomly select an individual from the population, what is the probability the person would be an obese male?
20. Based on the above problem, if I randomly select an individual from the population, and I know the person is obese, what is the probability that the person is male?
21. Height in the male population is normally distributed with a mean height of 5–11 (71 in.) and a standard deviation of 4 in. I am 5–8 (68 in.). In what percentile of the height distribution do I fall?
22. I like to flip coins, so I flip a quarter 100 times and obtain 55 heads. What is the probability, assuming the coin is fair, that I would obtain 55 or more heads in 100 flips of a fair coin?
23. A deck of cards consists of four "suits" (clubs, diamonds, hearts, and spades, with clubs and spades being black, and diamonds and hearts being red), each of which has 13 cards, including numbered cards from 2 through 10, three "face" cards (Jack, Queen, and King), and an ace, which can be considered a one or a high card above the King. In five card stud poker, players are each dealt five cards, and they must make the best hand possible out of the cards dealt. How many unique five card stud poker hands are possible?
24. If I deal you five cards, what is the probability of obtaining a royal flush? (royal flush is defined by having the 10, Jack, Queen, King, and Ace of a single suit).

25. What is the probability of obtaining a flush? (all cards are of the same suit. Don't exclude royal and straight flushes).
26. What is the probability that you are dealt all red OR all black cards?
27. What is the probability that you are dealt a four-of-a-kind (all four of a given number or face card)?
28. If I roll two dice, what is the probability that at least one of them will show a number greater than 4?
29. The breakfast buffet I went to last week had an omelet station with ham, green peppers, onions, cheese, bacon, mushrooms, and spinach as possible toppings. Assuming I can have as many of these toppings as I want (no repeats), how many different omelets are possible?
30. What is the probability that two people in a row will order the same omelet (assume they do not know each other and are unaware of the other's order)?
31. I flipped a fair coin 10 times in a row and got heads on every flip. What is the probability the next toss will land on heads?
32. What is the probability of obtaining 5 heads in a row on 5 flips of a fair coin?
33. There is a probability of .6 that I will give a tough final exam. If the final is easy, there is a probability of .8 that you will make an A. If the final is tough, there is a probability of .3 that you will make an A. What is the probability that you will make an A?
34. You find out you received an A. What is the probability that the exam was tough?
35. What is the probability of obtaining exactly 5 sums that exceed 8 on 10 rolls of a pair of dice?
36. If I draw two cards from a deck, what is the probability that I will obtain a four on the first draw followed by an ace on the second draw?
37. My local hardware store has a supply of 100 light bulbs, of which 5 are defective. If I buy two bulbs, what is the probability that both of them will be defective?
38. A certain machine has a triple redundancy system to minimize its probability of failure. If a particular component fails, a second one takes over the task. If the second one fails, the third takes over the task. If that component fails, then the machine fails. There is a probability of .8 that the first component will fail. If the second component is called, there is a probability of .5 that it will fail. If the second component fails, there is a probability of .3 that the third will fail. What is the probability that the machine will actually work?
39. I want to arrange 5 people out of a class of 10 on one side of a table. How many ways can I do this?
40. The probability that a Republican will win the presidential election is 0 if he does not win the state of Ohio. The probability that he will win if he does win Ohio is .35. If I tell you that the Republican won the election, what is the probability that he won Ohio?
41. The probability that a person will become rich if s/he does not finish college is (1/10,000). The probability that a person will become rich if s/he does finish

college is (1/10). The probability of college completion is .3. I introduce you to a rich person. What is the probability that she finished college?

42. In a Major League Baseball season, each team plays 162 games. Suppose a team's probability of winning each game is .5. What is the probability that a team will win more than 90 games?
43. What is the probability of rolling doubles (matching numbers) on a pair of dice?
44. I have three large koi in my outdoor pond. I'm hoping that they might produce baby fish, but they cannot do this if they're all the same sex (exclude the possibility of parthenogensis, which is known to happen in some fish). What is the probability that they're all the same sex? (assume the probability of .5 for each sex).
45. I'm trying to determine the proportion of persons in a community who tested HIV-positive in a recent community-wide blood test. This is a highly sensitive question, and many may not answer it directly. So, I ask the survey respondents to roll a die (and not show me the result). If the roll comes up 1 or 2, they are to answer "yes" to the question, regardless of whether they have HIV. If the die roll comes up 3 through 6, they are to answer truthfully. After collecting my data, I find that 50 % of respondents answers "yes" to the question. What proportion of the community in fact has HIV? (assume honesty, given the indirect method used, and assume the HIV test is 100 % accurate)
46. What is the probability that a person who said "yes" actually is HIV positive?
47. Assume that final exam scores in a given course are approximately normally distributed with a mean of 75 and a standard deviation of 10 points. In a class of 200 students, how many would you expect to fail the exam (i.e., score less than 60) (note: scores can only be approximately normally distributed, because the minimum score is 0, but this should not alter the answer, given how large the mean is relative to the standard deviation).
48. How many students can be expected to make an A; i.e., score 90 or higher?
49. The length of life of a certain brand of light bulb is normally distributed with a mean of 1,000 h and a standard deviation of 200 h (note: this is a very unreliable set of light bulbs!). If I buy new light bulbs for two lamps before I leave for a month-long vacation (30 days), and I leave both lamps on when I leave, what is the probability that at least one will be working when I return?
50. As described in a previous problem, height in the male population is normally distributed with a mean height of 5–11 (71 in.) and a standard deviation of 4 in. If I randomly selected 5 men from the population, what is the probability that I would have a collection of 5 men over 6 ft tall?

Chapter 6
Statistical Inference

In the last chapter, we developed probability theory and introduced the concept of mass and density functions, which can be used to handle large and/or complex sample spaces in which events have unequal probabilities of occurrence. These algebraic functions involve parameters and events (random variables), and if you know the parameters, you can deduce (compute) the probabilities for particular events—values of the random variable. Statistical inference involves reversing this process. When we take a sample from the population, we have a collection of events, but we usually do not know the values of the parameters that produced the observed sample data. The Central Limit Theorem plays an important role in helping us use sample statistics, like the sample mean and variance, to estimate population parameters, and especially to quantify uncertainty in our estimates.

6.1 The Central Limit Theorem and Inferential Statistics

In the previous chapter, we discussed how, if you assume the distribution for a variable in the population is normal, then we can use a z table to compute probabilities of obtaining a sample member with a value on the variable within any given range. The first step in the process of conducting statistical inference is to extend this idea of determining probabilities of obtaining a single sample member within a particular range of values on the variable to determining probabilities of obtaining an entire *sample* of a given size with a given value of a *statistic* like the mean from the population. For example, if we know the population mean, μ, is 10, we might want to know the probability of obtaining a sample of size $n = 100$ that has a mean, $\bar{x}$, of 8 or less.

In order to understand this process, we must discuss properties of statistics (like the mean) that can be drawn from a population. One of the most important theorems in statistics, the Central Limit Theorem (CLT), states that, as sample sizes increase, regardless of the distribution of a random variable in the population, sample means ($\bar{x}$'s) that can be drawn from that population distribution become normally

S.M. Lynch, *Using Statistics in Social Research: A Concise Approach*,
DOI 10.1007/978-1-4614-8573-5_6, © Springer Science+Business Media New York 2013

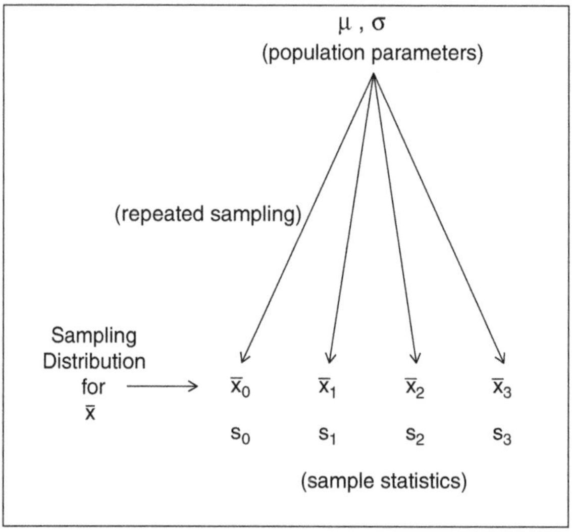

Fig. 6.1. Depiction of repeated sampling from the population and the concept of a sampling distribution.

distributed with a mean equal to the population mean (μ) and a variance equal to the variance in the population divided by the sample size used to compute the mean (σ^2/n). Equivalently, the standard deviation of the distribution of sample means is equal to $\sigma/\sqrt{n}$. The theorem's key result can be expressed in an abbreviated form as:

$$\sqrt{n}(\bar{x} - \mu) \xrightarrow{d} N(0, \sigma^2). \tag{6.1}$$

That is, the difference between sample means and the population mean, multiplied by the square root of the sample size, tends in distribution toward a normal distribution with a mean of 0 and a variance of σ^2. Another way to express the same idea is:

$$\bar{x} \overset{asy}{\sim} N\left(\mu, \frac{\sigma^2}{n}\right). \tag{6.2}$$

In this form, "asy" means asymptotically. That is, the distribution of sample means becomes more normally distributed as the sample size increases.

This theorem is conceptually difficult but forms the basis for statistical inference, and so some extended explanation is warranted. When you take a sample from a population, instead of thinking of this sample as a collection of individuals, consider that you are selecting a *sample mean* from the population. Figure 6.1 shows this process graphically. The population parameters μ and σ (or σ^2) govern the samples that we can draw from the population. If we take repeated samples (of a given size n) from the population and compute sample statistics, like $\bar{x}$, for each one, the CLT

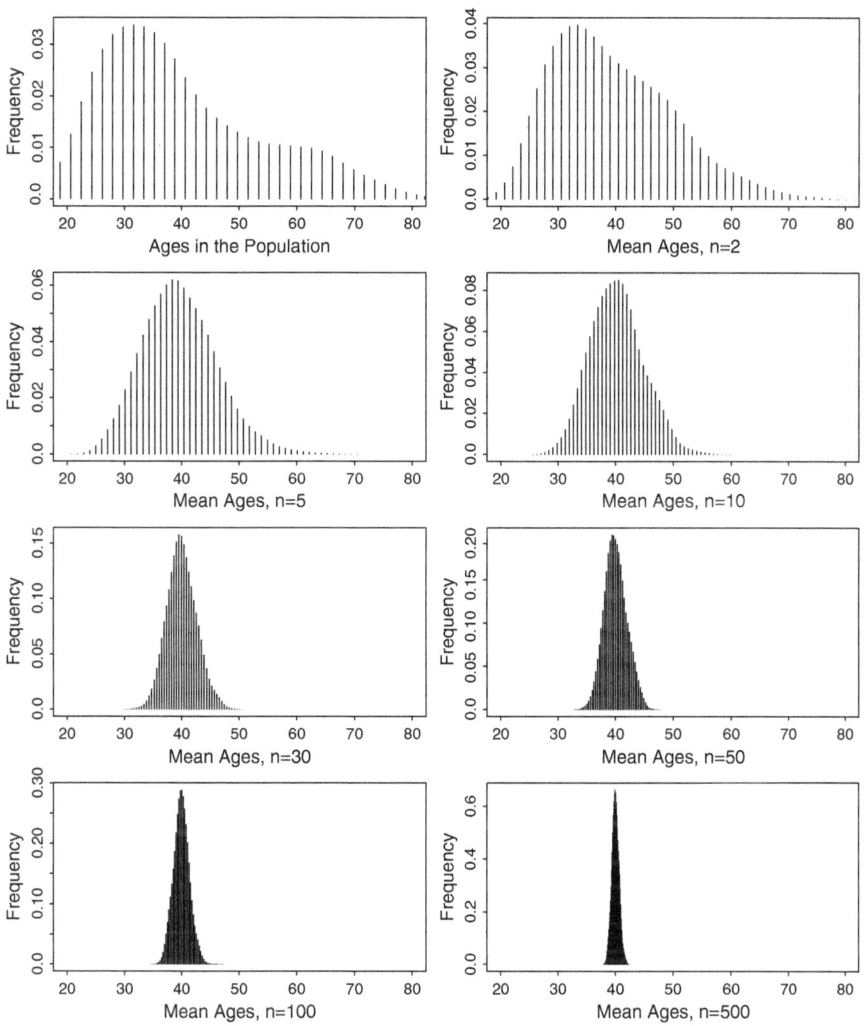

Fig. 6.2. Histograms of sampling distributions of the mean for various sample sizes

says that the distribution of these sample means—called the sampling distribution of the mean—will be normal with a mean of μ and a standard deviation of $\sigma/\sqrt{n}$, so long as n is "large."

In order to make these ideas concrete, I treated data on age from the 2004 General Social Survey as if it were a population distribution of ages, and from it I drew 1,000 random samples each of sizes n = 2, n = 5, n = 10, n = 30, n = 50, n = 100, and n = 500. For each of the samples, I computed the mean. So, for samples of size n = 2, I computed 1,000 means, for samples of size n = 5, I computed 1,000 means, etc. Figure 6.2 shows histograms of these collections of means by sample size.

Sample Size	$\mu_{\bar{x}}$	Observed s.e. ($\sigma_{\bar{x}}$)	Theoretical s.e. ($\sigma/\sqrt{n}$)
(Actual Population)	40.0	14.0 (σ)	14.0
2	39.4	10.2	9.9
5	39.7	6.3	6.3
10	40.2	4.5	4.4
30	40.0	2.5	2.6
50	40.0	1.9	2.0
100	40.0	1.4	1.4
500	40.0	.6	.6

Table 6.1. Results of simulation examining distributions of sample means

The upper left plot is the histogram of the original distribution of age. The upper right plot is the histogram of 1,000 sample means—called the sampling distribution—for the samples of size $n = 2$. The sampling distribution for the mean when $n = 2$ looks much like the original population distribution: it is skewed strongly to the right. However, as the figure shows, as the sample size increases, the sampling distributions become much more symmetric (normal), and their variance decreases. Ultimately, when the sample size is $n = 500$, the sampling distribution is very narrow.

Table 6.1 shows the means of the sampling distributions ($\mu_{\bar{x}}$), as well as the observed standard deviations of these sampling distributions ($\sigma_{\bar{x}}$; called the "standard error"—abbreviated s.e.). The far right column shows the theoretical standard error based on the CLT ($\sigma/\sqrt{n}$). Notice that the means of the sampling distributions are very close to the mean of age in the population, and that, when the sample size is 30, the mean of the sampling distribution is consistently within rounding of the true mean. Also notice that the observed s.e. consistently matches the theoretical s.e. ($\sigma/\sqrt{n}$) when the sample size is 100 or larger.

Why do the means of the sampling distribution become normal as the sample size increases, and why does the variance of the sampling distribution decrease? If you recall from the last chapter, the joint probability of two independent events is the product of their respective probabilities. When we take a simple random sample from the population, we are taking a collection of independent draws from the population distribution. Thus, their joint probability can be computed as the product of each of their probabilities. When the sample size is small, it may not be uncommon to draw a few extremely rare observations—observations that are far from the population mean—and thus obtain a sample mean that is far from the population mean. For example, the probability of obtaining two individuals whose values are rare enough that their probability of occurrence is .1 each is $.1 \times .1 = .01$. This is a small probability, but not incredibly small. However, it is extremely unlikely, if we draw a large sample, that we would draw a series of very rare values. For example, the probability of drawing five rare people like the ones we just discussed is $.1 \times .1 \times .1 \times .1 \times .1 = .00001$. This is an incredibly small probability. The implication is that it will be unlikely in a large sample to draw a large number of very rare individuals, and so our sample mean will tend to be close to the true mean.

The implication of the CLT for making inference about population means using sample means is that, if the sample size is large enough, we can very accurately and precisely estimate the population mean, and we can quantify our uncertainty in our estimate—that is, we can state how far away from the true population mean (μ) our estimate ($\bar{x}$) is likely to be.

6.2 Hypothesis Testing Using the z Distribution

Given that we know that the sampling distribution for $\bar{x}$ is normal, and we know its standard deviation (more on whether we "know" this later), we can construct a standardized score to determine the probability of obtaining a sample mean in some range from a sample of size n, given a particular true population mean:

$$z = \frac{\bar{x} - \mu}{\sigma/\sqrt{n}}. \tag{6.3}$$

Thus, if we knew the population mean were, say, 50, and the population standard deviation were 20, we could determine the probability of obtaining a sample of size 100 with a mean of $\bar{x} = 54$ or greater. The z-score would be:

$$z = \frac{54 - 50}{20/\sqrt{100}} = 2. \tag{6.4}$$

The corresponding probability of obtaining a sample with a mean of 54 or greater, then, would be roughly .025. In other words, it would be somewhat unlikely to obtain such a sample.

We can construct a similar z score if we are interested in computing the probability of obtaining a particular range for a proportion. For example, suppose we know that the proportion of persons in the population who support a particular policy is .6 (60 %), What is the probability of obtaining a sample of 500 persons in which less than 50 % of the respondents support the policy? In this situation, we simply replace μ with p_0 (the population proportion), $\bar{x}$ with $\hat{p}$ (the sample proportion), and σ with $\sqrt{p_0(1 - p_0)}$ (the standard deviation of a proportion):

$$z = \frac{\hat{p} - p_0}{\sqrt{p_0(1 - p_0)/n}} \tag{6.5}$$

$$= \frac{.5 - .6}{\sqrt{.6(.4)/500}} \tag{6.6}$$

$$= -4.56 \tag{6.7}$$

Thus, the probability of obtaining a sample of 500 persons with a proportion of .5 or smaller when the population proportion is .6 is $p(z < -4.56) \approx 0$.

Formal classical hypothesis testing utilizes this approach, but instead of using a known value for μ (which, in practice, we never have!), we substitute H_0 for μ into the calculation, where H_0 is called the "null hypothesis" we are interested in testing—it is a hypothesized value for the true population mean, μ. We then compute the probability of obtaining a sample mean as extreme as, or more than, our observed sample mean under this hypothesis. This probability is called a "p-value." If a sample mean is highly improbable to occur (i.e., p is small) under H_0, we reject H_0 and conclude that H_0 is probably not the true value of μ. It is common to say, when rejecting a null hypothesis, that the difference between the (null) hypothesized value for μ and the one potentially implied by the sample data, is "statistically significant" at the value at which p is considered to be small enough to reject the null. It is essential to recognize that the p-value *does not tell us the probability that H_0 is true.* It tells us the probability of obtaining our sample mean under the assumption that H_0 is true. In other words, hypothesis testing follows a "modus tollens" structure:

1. If H_0 is true, then $\bar{x}$ will be close to H_0 (its probability of occurrence will be high).
2. $\bar{x}$ is not close to H_0 (p is small).
3. Therefore, H_0 is not true (rejected).

A few questions/comments about this process are in order. First, how improbable must a sample mean be under the null hypothesis for us to reject the hypothesis? Second, what does this process tell us about the true value of the population mean?

Regarding the first question, there are two types of errors that can be made when following this hypothesis testing strategy: Type I and Type II errors. A Type I error is committed when we reject a null hypothesis that happens to be true. A Type II error is committed when we fail to reject a null hypothesis that is false. In practice, scientists are generally conservative and therefore most concerned with rejecting null hypotheses that are in fact true. Why? Suppose we were examining whether some experimental drug treatment is efficacious. The null hypothesis would be that there is no difference between the treatment and control groups on the outcome of interest. Rejecting this null would lend support to the view that the drug works, and we would want to be extremely confident that the treatment actually had an effect before marketing the drug, especially if it had negative side effects. Similarly, in social science, we want to be certain that socioeconomic, racial, sex, or some other difference in some outcome of interest in fact exists before we begin making policy recommendations that cost taxpayers money or cause some conflict.

The probability of making a Type I error is represented as α, and we generally set $\alpha = .05$ (called the "critical alpha") as our acceptable probability for making such an error in social research. This means that, when we conduct our hypothesis test, we want there to be a probability of less than .05 that we would observe a sample mean as extreme as (or more so than) ours if the null hypothesis were true (i.e., we want $p < \alpha = .05$). We generally split this probability across the two tails of the normal distribution, and so we typically will not reject a null hypothesis unless the z score that produces the p-value is greater than 1.96 (or less than -1.96). Thus, in the example above, if 50 were our hypothesized population mean, we determined that the probability of obtaining a sample with a mean at least as extreme as we had was .05. Extremeness is measured on *both* sides of the distribution, and so, $p(z > 1.96) + p(z < -1.96) < .05$; alternatively: $p(|z| > 1.96) < .05$. We call such an approach a "two-tailed test." We would therefore reject the null hypothesis and conclude that the population mean is probably not 50. The data simply aren't very consistent with that null hypothesis.

If we decided to set α much lower, it is possible that we might fail to reject a null hypothesis that would be rejected if α had been .05. In that case, the result would not be statistically significant at the chosen α level. Thus, we must keep in mind that statistical significance is somewhat subjective—it depends on the value chosen for α as much as it depends on the data. Furthermore, the fact that some result is declared to be statistically significant does not necessarily indicate that the result is substantively important. Consider the denominator of the z calculation: it depends on n. The larger the sample size, the larger z will be, even if the difference between $\bar{x}$ and H_0 remains constant in the numerator. As the CLT simulation showed, with a large enough sample size, it becomes extremely difficult to obtain a sample mean very different from the true population mean. Thus, with large n, even substantively trivial differences between $\bar{x}$ and H_0 may be declared statistically significant. For example, suppose our null hypothesis is that 80 % of people in the world like their statistics class (i.e., $p_0 = .8$). We obtain a sample of $n = 5,000$ persons who have taken statistics courses and find the 82 % liked their course. What would we conclude about the null hypothesis?

$$z = \frac{.82 - .8}{\sqrt{.8(.2)/5000}} = 3.53. \quad (6.8)$$

$p(|z| > 3.53) \approx 0$, and so we would reject the null hypothesis. But is the difference between our observed 82 % and our hypothesized 80 % worth discussing? Clearly, the vast majority of students like their statistics class.

This brings us to the second question: What does the process of hypothesis testing tell us about the true population mean? Unfortunately, the answer is: by itself, not much. It only tells us what the true population mean probably is not; it does not give us any indication what it is. The sample mean itself gives us an indication of this, however, and we will discuss this issue later in the chapter in the context of confidence intervals.

6.3 Hypothesis Testing When σ Is Unknown: The *t* Distribution

The discussion of hypothesis testing in the previous section assumed that the true population standard deviation (σ) is known. However, it is unreasonable to expect, if we don't and can't know μ, that we could possibly know σ. Fortunately, just as the CLT justifies our belief that $\bar{x}$ in large samples is very close to μ, the CLT also justifies the belief that s (the sample standard deviation) is close to σ in large samples. Thus, when σ is unknown, we may consider using s as our estimate of σ in making our standardized score calculations. However, given that there is uncertainty in the representation of σ with s (i.e., we don't in fact know how close s is to σ), the standardized score is no longer normally distributed. Instead, it is t distributed, and the associated score is thus no longer called a z score—it is called a t score:

$$t = \frac{\bar{x} - H_0}{s/\sqrt{n}} \tag{6.9}$$

The t distribution looks similar to the normal distribution. It is symmetric around its mean, median, and mode and is bell-shaped. However, it has fatter tails that reflect our uncertainty in using s as an estimate of σ. The probability density function for the t distribution is more complicated than the normal distribution, and it is unnecessary to discuss it. However, it is important to know what parameters are associated with the distribution. The t distribution has a mean and variance, just like the normal distribution, but it also has a "degrees of freedom" parameter (d.f.). For our purposes, the degrees of freedom associated with the distribution is $n-1$, where n is the sample size. As $n - 1$ gets larger, the distribution becomes more and more normal in appearance, reflecting the fact that a large sample size reduces uncertainty associated with using s to approximate σ. When d.f. > 120, the t and standard normal distributions are virtually indistinguishable. Figure 6.3 shows the t distribution with different degrees of freedom. Notice that, as the degrees of freedom increases, the tails of the t distribution flatten, and the distribution becomes more peaked in the center like the normal distribution. At 120 degrees of freedom, the figure shows that the t distribution cannot be differentiated from the z distribution.

6.4 Confidence Intervals

The process of hypothesis testing allows us to decide whether we think a hypothesized value for the population mean is reasonable, but it does not allow us to directly make inference about the true value of a population mean. In other words, the logic of hypothesis testing is that we computed the probability of observing the sample mean we did under some hypothesized value for the population mean. If

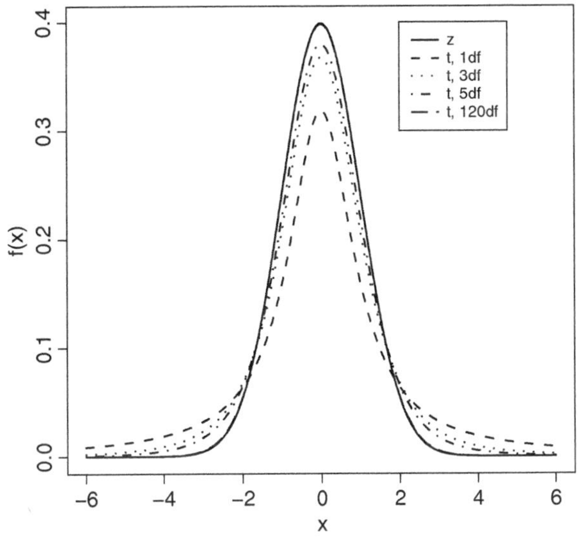

Fig. 6.3. Some t distributions

that probability (the p-value) is low, then we reject the hypothesized value. This approach does nothing to help us know, if the hypothesized value is rejected, what the true value of the population mean is. Confidence interval construction can, however, help us make inference about the population mean.

Confidence intervals are reported regularly in the popular media in discussions of both science and politics. During election seasons in particular, news agencies routinely report poll results describing preferences for particular candidates. For example, we might hear something like: "56 % of respondents favor candidate A over candidate B, with a margin of error of ±3 %". In this example, the polling agency is claiming that the true proportion of the population that favors candidate A is between 53 and 59 %. This result is a basic confidence interval.

The logic of confidence interval construction is quite simple. If we let the sample mean be our best guess for the population mean, and we let the sample standard deviation be our best guess for the population standard deviation, then, based on the CLT, we can construct an interval around our sample mean using the sample standard deviation and make some statement about where the true population mean is expected to fall.

Constructing confidence intervals follows the following steps:

1. Decide on a "confidence level." Just as in hypothesis testing, we generally establish $\alpha = .05$, which corresponds to a confidence level of .95.
2. Find $t_{\alpha/2}$. That is, find the value of t that leaves half of α in each tail of the t distribution.
3. Compute the standard error of the mean, which is $\frac{s}{\sqrt{n}}$
4. Construct the interval estimate as: $\bar{x} \pm (t_{\alpha/2} \times s/\sqrt{n})$

In the first step, we decide how confident we wish to be with respect to whether our estimated interval captures the population mean. Here we have to make a trade-off between precision and accuracy: the more precise we are, the less confident, and hence the less likely to be accurate, our estimate will be. If we increase our confidence (make $(1 - \alpha)$ large), we will necessarily have to increase our interval width. Conversely, if we decrease our confidence, we can narrow the width of the interval (it will then be less likely to capture the mean). Put in the extreme: we can be 100 % confident that the population mean falls in the interval $(-\infty\ ,\ \infty)$, or we can be 0 % confident that the population mean is exactly some number M. As a matter of convention, it is common in social science to choose $\alpha = .05$, which corresponds to a confidence level of 95 %.

In the second step, we find the appropriate t value that corresponds to the two-tailed probability $\alpha/2$ we are interested in. For example, if we want to be 95 % confident, then $\alpha = .05$, and we need to find $t_{.025}$—the value of t for which .025 % of the mass of the t distribution falls beyond it. Recall that the t distribution has a degrees of freedom parameter associated with it: if we have d.f. > 120, then the appropriate t value for this example would be 1.96 (same as the z). Note that, if σ is known, t can be replaced with z.

In the third step, we compute the standard error of the mean. This value, $s/\sqrt{n}$, derives from the CLT as we discussed before. It is a measure of the extent to which sample means derived from samples of a specific size (n) can be expected to vary from the true population mean. The factor $(t_{\alpha/2} \times s/\sqrt{n})$ is the margin of error.

Finally, we construct the interval estimate as indicated. One way to understand this interval estimate, and to relate it back to hypothesis testing, is to view it as a rearrangement, in a sense, of the t test we developed in the last section:

$$t = \frac{\bar{x} - \mu}{s/\sqrt{n}}. \tag{6.10}$$

If we multiply both sides by the denominator, we get:

$$t \times \frac{s}{\sqrt{n}} = \bar{x} - \mu. \tag{6.11}$$

If we then isolate μ:

$$\mu = \bar{x} - t \times \frac{s}{\sqrt{n}}. \tag{6.12}$$

This result looks remarkably like the interval estimate presented in step 4 above, with a couple of differences. First, the t statistic is computed under the hypothesis testing approach, but under the interval estimate approach it is fixed at a given value reflecting the confidence level (hence t is replaced with $t_{\alpha/2}$). Second, the interval estimate approach constructs an interval around $\bar{x}$, rather than simply a one-sided estimate (hence the "–" is replaced by "±"). Recall, however, that, when we obtain

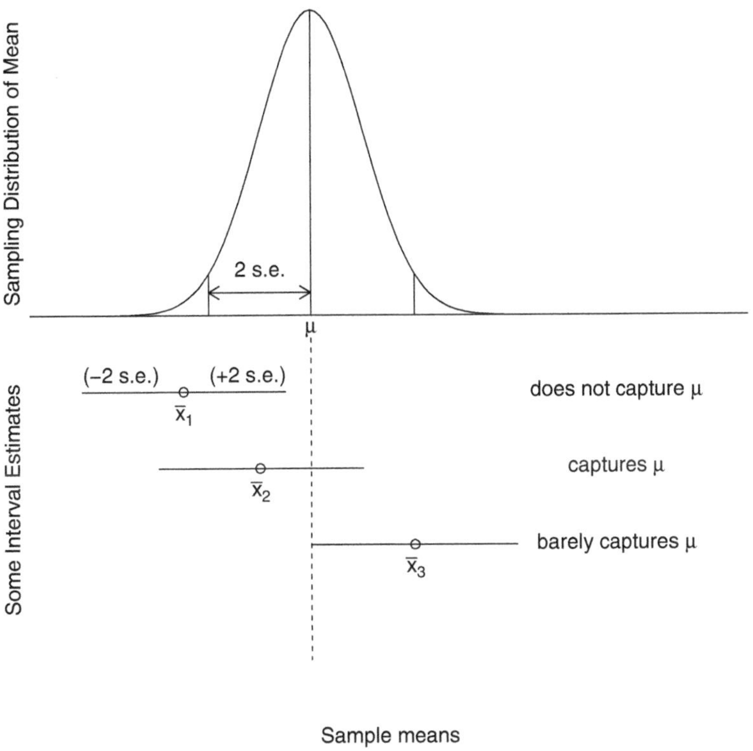

Fig. 6.4. Illustration of concept of "confidence." Plot shows that 95 % of the sample means that can be drawn from their sampling distribution will contain μ within an interval ranging 2 standard errors around $\bar{x}$.

a p-value (or choose an α and find a critical t), we explicitly assume that our t statistic could be positive or negative, which implicitly corresponds to using $\pm$ in the numerator of the equation for t.

The first difference between the hypothesis testing and interval construction approaches highlights the fundamental difference between hypothesis tests and intervals. Under hypothesis testing, we determine how probable a particular sample mean is under a hypothesized value of μ. Under the interval estimate approach, we predetermine the probability that intervals of a given width will contain the true value of μ.

How do we interpret confidence intervals? Ultimately, what we can say about our interval estimate is that, if we took repeated samples from the population and constructed $(1-\alpha)\%$ intervals around all of our sample estimates, $(1-\alpha)\%$ of such intervals would contain the true value of the parameter. Figure 6.4 illustrates this idea. The top half of the figure shows the sampling distribution for means $\bar{x}$ that can be drawn from a population with mean μ. Under the CLT, we know that this sampling distribution is normal, and so 95 % of all sample means (of a given size,

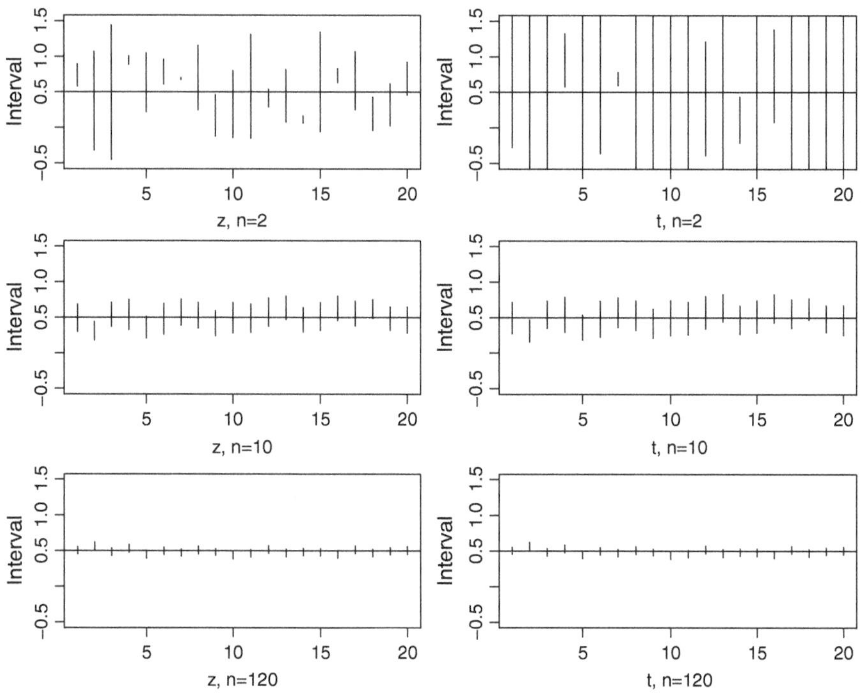

Fig. 6.5. Some confidence intervals from samples of different sizes (based on the z and t distributions)

n) that could be drawn from the population will fall within two standard errors of μ. Thus, if we take sample means and attach (add and subtract) two standard errors to them, 95 % of such intervals will contain μ. The bottom half of the figure shows three such intervals. In the first one, $\bar{x}_1$ is more than two standard errors away from μ, and so its interval does not contain μ. In the latter two, the sample means are within two standard errors of μ, and so, when we attach two standard errors to the sample estimates, the intervals contain μ.

Note that we cannot say, under this approach to constructing intervals, that μ falls within the interval with probability $1-\alpha$. We can only say that $(1-\alpha)\%$ of all intervals of a given width constructed in this fashion will contain μ. μ is considered fixed; only the intervals are random; they are the result of random sampling from the population. In order to demonstrate this idea, Fig. 6.5 shows some confidence intervals that were constructed around sample means which were drawn from a population for which the true mean was .5. In constructing the intervals, I used both the z distribution (critical z of 1.96) and the t distribution to show how uncertainty in using s as an estimate of σ diminishes as n increases. The plots in the left column of the figure are based on the z distribution; the plots in the right column are based on the t distribution.

Sample Size	z intervals	t intervals
n = 2	.67	.94
n = 10	.93	.95
n = 120	.94	.94

Table 6.2. Proportion of "95 %" confidence interval estimates capturing the true mean by type of interval (z/t) and sample size.

Notice that (1) not all of the intervals actually capture the true mean, (2) the proportion of intervals that do capture the true mean appears to increase as the sample sizes increase (ultimately, the proportion is limited to 95 %, because they are intervals based on $\alpha = .05$), and (3) the widths of the intervals in the smallest sized samples fluctuate drastically. This latter issue exemplifies why z-based intervals are not appropriate when the population standard deviation is unknown and the sample size is small: a large proportion of these intervals do not contain the population mean. However, when the intervals are based on the t distribution, they are wider and more capture μ.

Table 6.2 presents the proportions of confidence intervals (out of 1,000) that capture the true population mean for both z and t based samples of different sizes. Notice how the 95 % intervals based on the t distribution are almost always truly 95 % intervals; that is, that 95 % of them tend to capture the true mean, regardless of the sample size.

6.5 Additional Hypothesis Testing Tools

So far we have only discussed hypothesis testing and confidence interval construction for a single sample mean. Although this type of hypothesis test and confidence interval is important, we generally are more interested in comparing sample means—to test hypotheses about differences between groups—in actual social research. For example, we may be interested in determining whether men and women earn comparable incomes on average. Or, we may want to know whether whites and nonwhites vote similarly in an election. Or, we may want to know whether persons with a high school diploma are more likely to be depressed than persons with a college diploma. Or, we may want to know whether the proportion of persons with a high school diploma who are unemployed is comparable to the proportion of persons with a college diploma who are unemployed. These cases require extensions of our basic one-sample t test, and a number of extensions are available.

Although the tests we will discuss appear different from the one-sample test, they all derive from the basic one-sample t test and the basic logic of hypothesis testing. Suppose we were interested in the first question—whether men and women earn comparable incomes on average. We could hypothesize some difference between men and women in the population and construct the following t test based on the observed difference in sample means:

$$t = \frac{(\bar{x}_{men} - \bar{x}_{women}) - H_0 : (\mu_{men} - \mu_{women})}{S.E.(\bar{x}_{men} - \bar{x}_{women})}. \tag{6.13}$$

The test is possible because, under the CLT, both $\bar{x}_{men}$ and $\bar{x}_{women}$ are normally distributed, and the difference between two normally distributed variables is also normally distributed. What is the standard error here? This is the standard error of the difference between men and women. Recall that before, the denominator of the t statistic was $s/\sqrt{n}$. Here, we need an alternate measure of the sample standard error, because we are examining the difference between two groups. There are actually at least two ways to construct this standard error,[1] but the most common, and usually the most appropriate, uses a basic rule of variance algebra: the variance of the difference of two random variables is equal to the sum of the variances of the two random variables minus twice the covariance of the two random variables:

$$Var(A \pm B) = Var(A) + Var(B) \pm 2 \times Cov(A, B). \tag{6.14}$$

We will discuss the covariance later, but for now, note that two variables that are independent do not covary (i.e., they are uncorrelated). Because our sample members are selected independently, the mean for men and the mean for women are independent, and so the latter term involving the covariance is 0. Thus, we can compute the standard error of the difference between the means for men and women as:

$$S.E.(\text{difference}) = \sqrt{\frac{s^2_{men}}{n_{men}} + \frac{s^2_{women}}{n_{women}}}. \tag{6.15}$$

If we also recognize that, if we hypothesize a difference to be 0, then the latter two terms in the numerator disappear ($\mu_{men} - \mu_{women} = 0$), so our entire test reduces to:

$$t = \frac{\bar{x}_1 - \bar{x}_2}{\sqrt{\frac{s_1^2}{n_1} + \frac{s_2^2}{n_2}}}. \tag{6.16}$$

[1] A common and alternate calculation is $s_p\sqrt{\frac{1}{n_1} + \frac{1}{n_2}}$, where $s_p^2 = \frac{\sum(x_1 - \bar{x}_1)^2 + \sum(x_2 - \bar{x}_2)^2}{n_1 + n_2 - 2}$. This is called the pooled variance estimate. From a theoretical standpoint, s_p^2 is appropriate, if, members of the two groups come from the same distribution. However, we cannot really know this until *after* the test. Thus, I prefer the measure used in the text above, which assumes the samples are from different populations and makes the t-test more conservative (the non-pooled estimate is always at least as large as the pooled estimate). In the case the two groups are from the same overall population, the calculations are equivalent; if they aren't, the non-pooled estimate presented in the text is correct.

Here, I have simply replaced the "men" and "women" labels with 1 and 2 for generality. This t test is called the "independent samples t-test." The degrees of freedom associated with the test is tedious to compute and is as follows:

$$df = \frac{\left(\frac{s_1^2}{n_1} + \frac{s_2^2}{n_2}\right)^2}{\frac{s_1^4}{n_1^2(n_1-1)} + \frac{s_2^4}{n_2^2(n_2-1)}} \quad (6.17)$$

In most social science settings, the subsample sizes will be well over 120, and so the z distribution can be used as an approximation and the degrees of freedom calculation is not needed. Furthermore, when the subsample sizes do not exceed 120, it is usually the case that $df \geq \min(n_1-1, n_2-1)$. Thus, we can set the degrees of freedom equal to the smaller of the subsample sizes -1 for a conservative result.

As a simple example, suppose the mean income for men in a sample was found to be \$53,735, while the mean income for women was \$45,624. The standard deviations were \$32,508 and \$31,887, respectively, and the subsample sizes were $n = 12{,}038$ men and $n = 14{,}190$ women. If we construct the t statistic, we find:

$$t = \frac{53735 - 45624}{\sqrt{\frac{(32508)^2}{12038} + \frac{(31887)^2}{14190}}} = 20.31. \quad (6.18)$$

Thus, under the null hypothesis that the difference in incomes between men and women is 0, the probability of obtaining a sample in which the means were this different is $p(|t| \geq 20.31) \equiv p(t > 20.31) \times 2 \approx 0$. In other words, if mean income for men and women were in fact equal, we would practically never see this type of difference in a sample as large as we have.

Sometimes we may be interested in comparing groups, but in terms of comparing differences in proportions, rather than a continuous variable like income. In that case, we can simply replace our sample means with our sample proportions, and replace our sample standard errors with the standard errors for proportions, and the test proceeds as before. As we saw in Chap. 5, the variance of a proportion, p is $p(1-p)$. Thus,

$$t = \frac{\hat{p}_1 - \hat{p}_2}{\sqrt{\frac{\hat{p}_1(1-\hat{p}_1)}{n_1} + \frac{\hat{p}_2(1-\hat{p}_2)}{n_2}}}. \quad (6.19)$$

Here, I have used the symbol $\hat{p}$ to indicate that the sample proportion is an estimate of the population proportion p. I use this notation to emphasize that the variance of the proportions to be used is based on the sample proportions.

A final test that we may sometimes need is for examining pairs of individuals (or individuals' scores from two points in time), rather than comparing the means of independent samples. For example, suppose we want to determine whether some experimental treatment reduces depression, and suppose our data are individual depression measures taken before and after treatment. In this case, we do not have independent samples, because the two samples are really the same sample, just measured at different times. In that case, we simply construct the differences pairwise and compute the standard error of these differences. Thus, ultimately this test is a one-sample test on differences, rather than an independent samples test.

Table 6.3 summarizes all of these various z and t tests and confidence intervals we have discussed, in addition to a few that we have not. For example, we did not discuss the one sample z test for proportions, but it is a direct extension of the original one sample z test discussed earlier, just applied to proportions. It is not a t test, because, if p_0 (the population proportion) is hypothesized or known, then the population variance is known.

6.6 Conclusions

In this chapter, we discussed the Central Limit Theorem in some detail and developed the process of hypothesis testing and confidence interval construction from it. Hypothesis testing reveals the probability of obtaining the observed data under some hypothesis, allowing us to reject (or fail to reject) scientific hypotheses. However, it does not allow us to make inference about the true population parameter of interest. Confidence intervals do. A number of z and t tests and confidence interval strategies follow from the CLT and our original z test, and we will develop still more as we move to later chapters and topics.

6.7 Items for Review

- Central Limit Theorem
- Sampling distribution
- Null hypothesis and hypothesis testing
- p-value
- Type I and Type II errors
- α
- Critical value (for t or α)
- t distribution
- Degrees of freedom
- Margin of error
- Confidence interval
- Pooled and unpooled variance estimates
- Various statistical tests as shown in Table 6.3

Name	Formula
One-sample z test	$\frac{\bar{x}-H_0}{\sigma/\sqrt{n}}$
One-sample t test	$\frac{\bar{x}-H_0}{s/\sqrt{n}}$ df=$n-1$
Independent samples t test	$\frac{(\bar{x}_1-\bar{x}_2)-H_{\mu_1-\mu_2}}{\sqrt{(s_1^2/n_1)+(s_2^2/n_2)}}$ df=$\min(n_1-1, n_2-1)$
One sample z test for proportions	$\frac{\hat{p}-p_0}{\sqrt{p_0(1-p_0)/n}}$
Independent samples t test for proportions	$\frac{(\hat{p}_1-\hat{p}_2)-H_{p_1-p_2}}{\sqrt{(\hat{p}_1(1-\hat{p}_1)/n_1)+(\hat{p}_2(1-\hat{p}_2)/n_2)}}$ df=$\min(n_1-1, n_2-1)$
Paired Sample t test	$\frac{\bar{x}_{\text{diff}}-H_0}{s_{\text{diff}}/\sqrt{n}}$ df=$n-1$, where n is number of pairs
$(1-\alpha)\%$ Confidence interval for a mean	$\bar{x} \pm (t_{\alpha/2})(s/\sqrt{n})$ df=$n-1$
$(1-\alpha)\%$ Confidence interval for a proportion	$\hat{p} \pm (t_{\alpha/2})\left(\sqrt{\hat{p}(1-\hat{p})/n}\right)$ df=$n-1$
$(1-\alpha)\%$ Conf. Int. for difference in means	$(\bar{x}_1-\bar{x}_2) \pm (t_{\alpha/2})\left(\sqrt{s_1^2/n_1+s_2^2/n_2}\right)$ df=$\min(n_1-1, n_2-1)$
$(1-\alpha)\%$ Conf. Int. for difference in proportions	$(\hat{p}_1-\hat{p}_2) \pm (t_{\alpha/2})\left(\sqrt{\frac{\hat{p}_1(1-\hat{p}_1)}{n_1}+\frac{\hat{p}_2(1-\hat{p}_2)}{n_2}}\right)$ df=$\min(n_1-1, n_2-1)$

Table 6.3. Various hypothesis testing and confidence interval formulas following from the CLT.

6.8 Homework

1. Suppose years of schooling in the population has a mean of 12 and a standard deviation of 3. What is the probability of obtaining a sample of 50 people with a mean of 11 or more?
2. What is the probability that I could obtain a sample of 100 people in which the mean years of schooling was less than 12.1?

3. What is the probability of obtaining a sample of 500 people with a mean less than 11.8 or greater than 12.2?
4. If 52 % of the population supports a particular political candidate, what is the probability of obtaining a sample of 200 people in which less than 50 % of respondents support him/her?
5. I believe that systolic blood pressure averages 120 mmHg in the population with a standard deviation of 10 mmHg. If I randomly sampled 50 people and found that mean systolic blood pressure was 130, what could/would I say about my assumption that the population mean is 120?
6. If a sample mean for years of schooling was 13.58 with a standard deviation of 2.66 years (n = 2,386), and you wanted to be 95 % confident in estimating the population mean education level, what would you say (i.e., what would your interval estimate be)?
7. I've been told that one-third of the population is in excellent health. In the GSS, 32.4 % of the sample ($n = 26{,}228$) claims excellent health. Is what I've been told reasonable?
8. Does income predict happiness? In a subsample from the GSS, mean happiness for those with higher than average income is 1.32 (s = .59, n = 11,366), while mean happiness for those with lower than average income is 1.11 (s = .64, n = 14,862).
9. Construct 95 % confidence intervals for happiness by income group using the data in the previous question.
10. Do whites and blacks experience different levels of life satisfaction? In the GSS, a mean satisfaction scale score for whites is 23.03 (s = 4.52, n = 10,791) and for blacks is 20.55 (s = 5.07, n = 1,269).
11. I've heard that real incomes have been stagnant since 1973. In the GSS, mean income in 1973 was $53,641 (s = $31,213, n = 1,106). In 2006 (the last data available, say), mean income was $54,291 (s = $40,511, n = 1,420). Are the data consistent with that claim?
12. In a sample of 10 men, 70 % claim to like country music. In a sample of 8 women, 65 % of claim to like country music. Do men's and women's musical tastes differ?
13. Suppose I think that the country is, on average, politically moderate. In 2006, the GSS shows a mean party affiliation score of 3.32 (s = 1.71, n = 1,420). The range of the score is 1–6 (strong democrat... strong republican), so that the "middle" of the scale is 3.5. Do the data support this view that the country is moderate on average?
14. Some argue that the US population is aging. Is there any evidence of this in the GSS? In the GSS, the mean age of the sample was 43.89 in 1972 (s = 16.88, n = 1,086) and 46.10 in 2006 (s = 16.85, n = 1,420).
15. My theory says that men marry women with comparable levels of intelligence: that, on average, neither husbands nor wives are smarter than their partners. Assume I measure intelligence with IQ for 10 couples. What can I conclude about my hypothesis?

Couple	Husband's IQ	Wife's IQ
1	100	110
2	140	135
3	90	120
4	150	120
5	140	130
6	95	110
7	100	95
8	50	50
9	200	200
10	100	95

16. Some say that women are discriminated against in the labor market, that is, they do not receive comparable incomes for the same level of education as men. In a GSS subsample for men and women with comparable levels of education, mean male income was $47,828 with a standard deviation of $30,031. Mean female income was $41,354 with a standard deviation of $26,333. There were 277 men in the sample and 351 women in the sample. Is there evidence that men's and women's incomes at this education level differ?
17. Some political scientists argue that women and men differ in their political views. In a GSS subsample, among women, 487 claimed to be liberal, 448 claimed to be moderate, and 366 claimed to be conservative. Among men, 332 claimed to be liberal, 416 claimed to be moderate, and 337 claimed to be conservative. Is there any evidence that political views differ between sexes?
18. Exit polling of 400 voters in a district found that 55 % had voted for candidate A over candidate B. Given this information, would you call the election for candidate A?
19. My neighbor estimates that mean family income in the US population is $50,000. In our GSS sample of 2,386 persons, mean family income was $51,120.18 with a standard deviation of $32,045.84. If we assume that the GSS sample is random, do you think my neighbor has a reasonable estimate?
20. I want to be sure to obtain a margin of error of ± 2 % in a poll asking whether people support the government restricting the purchase of extra large carbonated beverages. How large should my sample be?
21. Based on anecdotal evidence (i.e., nonrandom, small samples), some have hypothesized that IQ has actually improved among recent birth cohorts. When a particular IQ test was validated in the 1950s, the mean was 100, with a standard deviation of 15. Suppose I hypothesize that IQ now has a mean of 110 (but with a standard deviation still of 15). I take a random sample of 20 persons and find $\bar{x} = 105$. What can I say about my hypothesis?

22. Now suppose I am unwilling to assume a standard deviation of 15, so I use my sample standard deviation of 12 instead. Reconduct the hypothesis test appropriately.
23. Explain why we may be more concerned about committing type I errors rather than type II errors.
24. It is a longstanding view that men tend to inflate the number of sexual partners they've had, while women tend to deflate their number. Part of the rationale for this view is that men need to appear "macho." I hypothesize that, by age 50, however, married men and women shouldn't differ in the number of sexual partners claimed. According to the GSS, the mean number of sexual partners claimed (in the last year) by 1886 married men over age 50 was 1.14 (sd = 2.58), while the mean number of sexual partners claimed by 1,750 married women over age 50 was .95 (sd = .44). Is my hypothesis reasonable?
25. Arguably, for a married person one would expect that the mean number of sexual partners reported would be 1. In the GSS, of 3,636 persons over age 50, the mean number of partners reported was 1.049 (sd = 1.89). Is it reasonable that the mean number of partners is 1 in this population?
26. Some argue that republican presidencies are worse for the economic condition of families in the U.S. than are democratic presidencies; others argue the opposite. In the GSS, 22.56 % of 30,979 respondents reported being in worse financial condition than the year before during GSS survey years in republican presidencies, while 20.42 % of 17,681 respondents reported being in worse financial condition than the year before during democratic presidencies. Is either party better, at least based on these data?
27. The term "bleeding heart liberal" is due to the (supposed) excess sympathy that liberals feel for criminals, the poor, the sick, and other marginalized groups. At the same time, however, members of marginalized groups are also more likely to be liberal, arguably because they themselves have experienced circumstances that make them more empathetic to the plight of others. If this hypothesis is true, we might expect that self-proclaimed liberals would have higher mean scores on an index capturing the number of "bad things" that have happened to them in the last year. In the GSS, of 56 persons claiming to be "strong conservatives," the mean score on such an index was 3.75 (sd = 4.15), while, of 46 persons claiming to be "strong liberals," the mean score on the index was 4.4 (sd = 2.30). Is the hypothesis reasonable?
28. An exit poll of 300 persons showed that candidate A received 53 % of the vote in a given precinct. If you were the pollster, would you call the election for candidate A based on these results?
29. In some states, if there is a difference of less than one percentage point between two candidates' final vote tallies, a recount of the cast ballots is automatically performed. In such a state, suppose that an exit poll of 200 people found 103 votes for one candidate and 97 for the other. Do you suspect there will be a recount?

30. I want to be sure to obtain a margin of error of ±5 % in a poll asking whether people support universal background checks for purchasing guns. How large should my sample be?
31. A recent survey of 1,000 people found that 90 % of Americans support background checks for purchasing guns. Construct a 95 % confidence interval for the population proportion.
32. A group of 10 persons participated in a weight loss study evaluating the efficacy of a new drug. Based on the data below, would you conclude that the drug worked?

Person	Before	After
1	300	280
2	280	285
3	425	360
4	315	330
5	255	249
6	600	590
7	290	300
8	265	235
9	310	290
10	230	250

33. Construct a 95 % confidence interval for the weight change in the previous problem.
34. In the 2010 GSS survey, 492 persons out of a total of 1,030 said that we spend too little on law enforcement in the US. I believe the country is evenly split on the issue. Am I right?
35. In the 2008 GSS, those who said they were raised as religious fundamentalists had a mean educational attainment of 13.09 years (sd = 2.93; n = 516), while those who were not raised as fundamentalists had a mean educational attainment of 13.64 years (sd = 3.28; n = 1,083). Is there a difference in educational attainment between fundamentalists and nonfundamentalists in the population?
36. A strip of the south has been called the "Bible Belt." Is there evidence that the region is distinct from other regions, in terms of the religiosity of its people? In the 2008 GSS, 275 out of 591 southerners claimed to have been raised as fundamentalists, while 242 out of 1,012 nonsoutherners were raised as fundamentalists.
37. Suppose I'm a firm believer that no statement is so outrageous that you can't get at least 10 % of the population to believe it. So, I conduct a poll of 200 people (randomly chosen) and ask whether they believe the following statement: "The earth was constructed by invisible, purple, one-legged unicorns all named Bill exactly 5,000 years ago this Thursday at noon." In my sample, 32 people said they had no doubt this was true. What would you conclude about my hypothesis? (assume, for the present purposes, that my statement is the most outrageous possible).

38. If 50 % of the population supports a national health care plan, what is the probability of obtaining a sample of 500 persons in which less than 45 % do?
39. Some say that support for the death penalty is declining. Using GSS data, I find that, in 2010, 1,297 respondents favored the death penalty, while 627 opposed it. In 1980, 982 respondents favored it, and 390 opposed it. Construct a 95 % confidence interval for the difference in support across these two time periods. Based on the evidence, would you agree that support has declined?
40. Capture-recapture methods are statistical methods that can be used to estimate the size of an unknown population. For example, suppose I am interested in estimating the total number, N, of fish in a pond. First, I catch a number of fish (say n_1 fish), tag them, and then release them (this is the "capture" step). Next, I go fishing again and catch n_2 fish (this is the "recapture" step). I observe that x of them are tagged. Using these data I can estimate the total number of fish in the pond as follows:

$$\frac{n_1}{N} = \frac{x}{n_2} \tag{6.20}$$

$$N = \frac{(n_1)(n_2)}{x} \tag{6.21}$$

In words, when I am finished with the capture step, the proportion of the total number of fish in the pond that are tagged is $n_1/N = p_0$. I can estimate this proportion via the recapture process: $x/n_2 = \hat{p}$. Since, on average, we would expect $\hat{p}$ to equal the population proportion p_0 (under the CLT), we can set these two proportions equal and solve for the only unknown, N.

Now, if I were to engage in a second recapture process, I would most likely obtain a different $\hat{p}$, and so it is easy to see that $\hat{p}$ can be viewed as the key random quantity here. N is constant, and both n_1 and n_2 can also be held constant. So, if we were interested in comparing the number of fish in two ponds, then, we really can just compare $\hat{p}$ across the two ponds.

So, suppose I run a catfish farm, and a fish food company claims that their food is not only cheaper than that of their competitors, but it also improves the life expectancy of fish, and keeps them healthier so that they can escape from predators (e.g., predatory birds). I decide to test their claim in a year-long experiment. I have two identical holding ponds into which I randomly distributed 1,000 hatchlings each. For pond A, I used my usual feed; for pond B, I used the new feed. A year later, I decide to estimate the total number of fish in each pond, with the hypothesis being that, if the fish food company's claim is true, pond B should have more fish remaining than pond A. I spend a day fishing each pond, capturing 100 fish in each pond, tagging them, and releasing them. A few days later, I spend a half day fishing. In pond A, I catch 48 fish, 6 of which are tagged. In pond B, I catch 62 fish, 7 of which are tagged. Using the methods that have been covered in this chapter, and the logic of capture-recapture methods as described above, what would you conclude about the fish food company's claim?

In reality, capture-recapture methods are a little more complicated than this in practice, and we need to make several assumptions to justify their use. What do you think some of those assumptions are, and what are some problems that you see with the method as described here? In other words, think critically about the method like a statistician might.

Chapter 7
Statistical Approaches for Nominal Data: Chi-Square Tests

In the previous chapter, we considered statistical tests that involve (1) a single continuous numeric variable, the sample mean for which is tested against a hypothetical population value, and (2) a single continuous numeric variable in which we compare means across a two category nominal level grouping variable like sex. Sometimes, however, we are faced with two nominal level variables we think may be related, like marital status and region of the country. In other cases, we may have a pair of ordinal variables, or a combination of nominal and ordinal variables, we expect to be associated but perhaps not linearly. For example, we may be interested in the relationship between religious affiliation (e.g., Protestant, Catholic, Jewish, Other, None) and educational degree attainment (e.g., high school diploma, 2-year college degree, 4-year college degree, more than a 4-year degree).

In these cases, the mean of one or both of the variables has no real meaning, and so tests that compare means are inappropriate. Additionally, even if comparing means *might* make some sense (e.g., examining *mean* educational attainment by religious affiliation), we may not expect a clear pattern that will be detectable as a difference in means. Table 7.1 presents some hypothetical religious affiliation and educational attainment data. In these data, there are 100 Catholics and 200 Protestants. All of the Catholics have a high school diploma; there is no variability in educational attainment. None of the Protestants has a high school diploma; instead, exactly half have less schooling, while half have more.

If we were to treat educational attainment as numeric, giving the categories values of 1, 2, and 3, respectively, mean attainment would be identical for the two religion groups. As a result, the numerator of an independent samples t test would be 0, and we would reject the null hypothesis that the means do not differ between the two groups. However, this finding is clearly not the end of the story, because it ignores a clear pattern in the data.

In cases like the one presented in this example, as well as the case in which both variables of interest are measured at the nominal level, we can turn to a "nonparametric" test, the most basic of which is the chi-square test of independence. The test is called "nonparametric" for two reasons. First, unlike the z and t tests discussed in the previous chapter, the chi-square test of independence does

S.M. Lynch, *Using Statistics in Social Research: A Concise Approach*,
DOI 10.1007/978-1-4614-8573-5_7, © Springer Science+Business Media New York 2013

	Religion		
Education	Catholic	Protestant	TOTAL
Less than High School	0	100	100
High School Diploma	100	0	100
More than HS Diploma	0	100	100
TOTAL	100	200	300

Table 7.1. Crosstabulation of hypothetical religious affiliation and educational attainment data (all religions not shown have been excluded).

	Marital Status				
Region	Married	Widowed	Divorced/ Separated	Never Married	TOTAL
Northeast	91	20	29	61	201
Midwest	174	30	71	79	354
South	246	30	115	120	511
West	174	30	60	90	354
TOTAL	685	110	275	350	1420

Table 7.2. Crosstabulation of marital status and region of the country (2006 GSS data).

not depend on parameters like the mean and variance. Second, the test does not represent, describe, or reveal patterns observed in data using other types of parameters like regression analysis does, as will be discussed in later chapters.

The starting point for the chi-square test is a crosstab of two variables. Thus, you may consider this test when you have data that you would summarize using a crosstab. Table 7.2 is a crosstab of region of the country by marital status in the 2006 GSS. As the table shows, there are 91 persons (out of the 1,420 in the 2006 sample) who are both married and live in the northeast, 20 who are widowed and live in the northeast, 29 who are divorced or separated and live in the northeast, and so on. In total, there are 201 persons in the sample who live in the northeast, 354 who live in the midwest, 511 who live in the south, and 354 who live in the west. Similarly, there are 685 who are married, 110 who are widowed, 275 who are divorced or separated, and 350 who have never been married.

Notice that both of these variables are measured at the nominal level; that is, there is no way to order the categories in any meaningful way. Also notice that the raw counts themselves vary substantially from cell to cell—ranging from 20 to 246—and the marginals also vary substantially. There are far more married persons (685) than widowed persons (110), and the region distribution shows that far more live in the south (511) than in the northeast (201).

All in all, given that the raw counts vary considerably from cell to cell and across the marginals, and given that the categories of each variable cannot be meaningfully ordered, any relationship that may exist between marital status and region is difficult to discern immediately from the raw cell counts. For this reason, as we did in Chap. 4, we generally construct and include row and/or column percentages in the crosstab, depending on which variable we consider to be the independent (predictor)

	Marital Status				
Region	Married	Widowed	Divorced/ Separated	Never Married	TOTAL
Northeast	91	20	29	61	201
	45.3 %	10.0 %	14.4 %	30.3 %	100 %
Midwest	174	30	71	79	354
	49.2 %	8.5 %	20.1 %	22.3 %	100 %
South	246	30	115	120	511
	48.1 %	5.9 %	22.5 %	23.5 %	100 %
West	174	30	60	90	354
	49.2 %	8.5 %	16.9 %	25.4 %	100 %
TOTAL	685	110	275	350	1420
	48.2 %	7.7 %	19.4 %	24.6 %	100 %

Table 7.3. Crosstabulation of marital status and region of the country (2006 GSS data) with row percentages included.

vs. dependent (outcome) variable. Here, our theory might suggest that regional culture may influence marital patterns. For example, the so-called "Bible Belt" runs through the south, and so it may be more likely for individuals to become and remain married than to be separated or divorced, or at least this pattern may be stronger in the south than in other regions.

To assist us in providing a descriptive depiction of this expected pattern, given the structure of our marginals (marital categories along the horizontal and regions along the vertical), we may incorporate row percentages in the table. Table 7.3 expands Table 7.2 to include these percentages. The table shows that our initial hypothesis may not be supported. Of those who live in the south, 48.1 % are married, while the percentages who are married of those who live in other regions are generally higher. For example, 49.2 % of those who live in either the midwest or the west are married. Only those who live in the northeast are less likely to be married (45.3 %).

Of course, one could argue that the age distribution of each region varies. So, the facts (1) that marital status transitions (e.g., going from being never married, to married, to divorced or separated or widowed) are age-patterned, and (2) that southerners may be either younger than persons in other regions (due to higher fertility levels) or older (due to migration to Florida post-retirement), may explain the lower percentage married. In that case, we might expect that southerners have a higher percentage who are either never married (the younger folks) or widowed (the older retirees). However, the comparisons of the percentages of persons in other marital status categories fail to support this view.

In terms of divorce or separation, the south has the highest percentage: 22.5 % of those who live in the south fall in this marital status category, while the proportions in this category range from 14.4 to 20.1 % in the other three regions. Furthermore, the south has the smallest percentage widowed and the second smallest percentage never married. Thus, the results suggest the age distribution argument is not a reasonable explanation. Instead, the results suggest that southerners divorce or separate at a high rate once they marry.

7.1 The Chi-Square (χ^2) Test of Independence

The region-by-marital status crosstab is informative, but there are limits to what can be determined from it. Most importantly, the data are sample data, and so we can't immediately conclude that patterns observed in the data exactly match those in the population. That is, we cannot immediately know whether the observed sample patterns are due to random fluctuation from sample to sample that could be drawn from a population in which there is no pattern, or whether they are real patterns. Less importantly, but also important, row and/or column percentages do not necessarily show us where the strongest patterns are in the data, especially once random variation is considered. For example, some 30 % of persons in the Northeast are in the "never married" category, while 22 % of those in the Midwest are in that category. At the same time, some 10 % of those in the Northeast are widowed, while 5.9 % of those in the South are widowed. Both of these differences in percentages seem to be substantively significant contributors to the overall pattern of regional differences in marital status. Yet, which difference is more important to the overall pattern, if one exists after considering sampling variability? The chi square test of independence can help us answer these questions.

The chi square test of independence begins with the assumption that the two variables of interest are independent in the population and involves computing the probability of observing the sample data under this assumption. As with the hypothesis testing approach outlined in the last chapter, if the probability of observing the sample data is small under the assumption of independence, we reject this "null" hypothesis and conclude that the variables are probably not independent in the population.

How do we compute the probability that the variables of interest are independent in the population? Recall from Chap. 5 that, if two events are independent, then their joint probability ($p(A, B)$) is the product of their respective marginal probabilities. In the context of the current example, one set of events is the region in which the respondent lives, and the other set is the marital status the respondent holds. Thus, "living in the Northeast" is an event, as is "being married," and so being married and living in the Northeast is a joint event. If these events are independent, then the probability that a randomly selected individual is both living in the Northeast and is married is simply the product of the probability that s/he lives in the Northeast with the probability that s/he is married.

The probability of living in the Northeast, given the data, can be computed from the row marginal: 201 of the 1,420 respondents live in the Northeast; thus, the probability of living in the Northeast is $201/1{,}420 = .142$. The probability of being married, given the data, can be computed from the column marginal: 685 of the 1,420 respondents are married; thus, the probability of being married is $685/1{,}420 = .482$. Therefore, the joint probability of living in the Northeast and being married, if region and marital status are independent, is $.142 \times .482 = .0684$.

Of a sample of $n = 1{,}420$ persons, this result means the number of persons that we *should* see in this category, under the assumption of independence, would be

	$X = 0$	$X = 1$	Total (Marginals for Y)
$Y = 0$	a	b	$a + b$
$Y = 1$	c	d	$c + d$
Total (Marginals for X)	$a + c$	$b + d$	$a + b + c + d = n$

Table 7.4. Generic cross-tab of variables X and Y

$E = 1{,}420 \times .0684 = 97.1$. However, the data show that 91 persons hold this joint status. The difference between the observed ("O") count and the expected count ("E") is a measure of the degree of discrepancy between reality and what reality would be expected to look like if region and marital status were independent.

Each cell in the crosstab provides us with an observed count. We can compute an expected count for each cell as we did for the first cell, and we should find that our expected counts sum to the overall sample size. Thus, a simple cell-by-cell sum of the difference between the observed and expected counts will be 0.

Computing the chi-square test statistic involves finding the discrepancies between observed and expected counts by cell, squaring them, dividing them by the expected count, and summing these transformed discrepancies across all cells. Mathematically, the chi-square statistic is computed as:

$$\chi^2 = \sum_{r=1}^{R} \sum_{c=1}^{C} \frac{(O_{rc} - E_{rc})^2}{E_{rc}}, \tag{7.1}$$

where O_{rc} is the observed cell count in row r and column c, E_{rc} is the expected cell count under the assumption of independence, and the summation is across all cells in the R by C table.

The generic computation of the expected cell counts can be simplified from what I presented above as:

$$E_{rc} = \frac{(O_r)(O_c)}{n}. \tag{7.2}$$

Why? Consider the generic crosstab in Table 7.4. If X and Y are independent, then $p(x, y) = p(x) \times p(y)$. So, the probability of falling in the $(0, 0)$ cell is:

$$P(x = 0, y = 0) = p(x = 0) \times p(y = 0) = \frac{a + c}{n} \times \frac{a + b}{n}. \tag{7.3}$$

Thus, $p(x = 0, y = 0) = \frac{(a+c)(a+b)}{n^2}$. The expected cell *count* , then is:

$$n \times p(x = 0, y = 0) = n \times \frac{(a + c)(a + b)}{n^2} = \frac{(O_{x=0})(O_{y=0})}{n}. \tag{7.4}$$

Once we have computed the expected cell counts under the assumption of independence, we need to determine how far from this assumption the observed cell counts are. This is captured by the test statistic formula shown above: $\sum_{r=1}^{R}\sum_{c=1}^{C}\frac{(O_{rc}-E_{rc})^2}{E_{rc}}$. This formula measures the sum of the squared differences between the observed and expected counts, after these squared differences have been adjusted for the magnitude of the expected counts. That is, if the expected count of observations is large, then we would probably expect the squared deviation between observed and expected to be large also, even if the variables are, in fact, independent. So, these squared deviations are adjusted somewhat for the magnitude of the expected cell count.

If we were to take repeated samples from a population in which two variables were independent, we would expect the sum of the squared deviations as computed above to vary across samples. Some samples would have greater sums of squared deviations than others. Overall, the sums across repeated samples would follow a chi-square distribution, just as sample means in repeated samples follow a normal distribution. So, we can use the chi-square distribution to assess the probability of observing our data under the independence assumption.

The chi square distribution is a probability distribution like the binomial, normal, and t distributions we have discussed before (see the tables at the end of the book for more discussion). It is a strictly non-negative distribution (sums of squares cannot be negative), and so it is often skewed to the right. As with the t distribution, the chi square distribution has a degree of freedom parameter that changes the shape of the distribution. For the chi square test of independence, the degrees of freedom are computed as: $(R-1)(C-1)$, where R and C are the numbers of rows and columns (respectively) in the crosstab table. This calculation simply represents the number of cells whose counts need to be known (in addition to the marginals) before the entire crosstab is determined. For example, in a two-by-two crosstab, the count in only one cell must be known before the counts in all remaining cells are determined.

7.1.1 *The Test Applied to the Region and Marital Status Data*

Table 7.5 shows the chi square test of independence calculations applied to the region/marital status data (only first two columns show $O_r O_c/n$ computations). The test statistic for these data was $\chi^2 = 14.65$ on $(4-1)(4-1) = 9$ degrees of freedom. If we look this chi square value up in Appendix A, we find that our χ^2 value is just short of reaching the $p < .1$ level, and it is certainly too small to be statistically significant at the usual $\alpha = .05$ level. Thus, we cannot reject the null hypothesis that region and marital status are independent. Notice that, unlike the z and t tests we conducted in the previous chapter, we only use one-tail p-values in the chi square test. The reason for this is that the distribution is non-negative, and perfectly independent variables should therefore produce a chi square value of 0. There is no left-tail extreme value to consider. Instead, the question is simply whether our observed discrepancies between the observed data and the expected data under the independence assumption are large enough to make us doubt the independence assumption.

	Marital Status				
Region	Married	Widowed	Divorced/ Separated	Never Married	TOTAL
NE	91	20	29	61	201
E:	$\frac{(201)(685)}{1420} = 97.0$	$\frac{(201)(110)}{1420} = 15.6$	38.9	49.5	
$\frac{(O-E)^2}{E}$	.37	1.24	2.52	2.67	6.80
MW	174	30	71	79	354
E:	$\frac{(354)(685)}{1420} = 170.8$	$\frac{(354)(110)}{1420} = 27.4$	68.6	87.3	
$\frac{(O-E)^2}{E}$	.06	.25	.08	.79	1.18
S	246	30	115	120	511
E:	$\frac{(511)(685)}{1420} = 246.5$	$\frac{(511)(110)}{1420} = 39.6$	99.0	126.0	
$\frac{(O-E)^2}{E}$	.001	2.33	2.59	.29	5.20
W	174	30	60	90	354
E:	$\frac{(354)(685)}{1420} = 170.8$	$\frac{(354)(110)}{1420} = 27.4$	68.6	87.3	
$\frac{(O-E)^2}{E}$	.06	.25	1.08	.08	1.47
TOTAL	685	110	275	350	1420

Table 7.5. Crosstabulation of marital status and region of the country (2006 GSS data) with chi square test of independence calculations ($\chi^2 = 14.65$, $df = 9$, $p > .10$)

7.2 The Lack-of-Fit χ^2 Test

The chi square test can be modified for the case in which one has either true or hypothesized population values and observed sample values and wishes to test whether the observed sample came from the known (or hypothetical) population. The computation of the test is the same as with the test of independence, but the data are generally arranged in a list. The degrees of freedom for the lack-of-fit test is $C - 1$, where C is the number of groups in the population.

As a simple example of the lack-of-fit test, suppose I know that the population in 2000 was 80 % white, 14 % black, and 6 % other races. Of the 1,667 persons in the 2000 sample of the GSS, 1,380 were white, 173 were black, and 114 were of other races. Given the known population proportions, in a sample of 1,667 persons, we would expect to have $1{,}667 \times .8 = 1{,}333.6$ whites, $1{,}667 \times .14 = 233.4$ blacks, and $1{,}667 \times .06 = 100.0$ others. Table 7.6 shows the observed, expected, and chi-square computations for these data and population proportions. The lack-of-fit statistic was 19.2 on 2 degrees of freedom. This value is much larger than 13.82, the critical value needed for obtaining $p < .001$. If our population proportions had been hypothesized, we would therefore reject the hypothesized values: the sample would probably not arise from the hypothesized proportions.

In this case, however, the population proportions were known. Given that the observed data would probably not occur in a simple random sample, given the

Race	Sample (Observed)	Expected (Pop. %×n)	$(O-E)^2/E$
White	1380	1333.6	1.61
Black	173	233.4	15.63
Other	114	100.0	1.96
TOTAL	1667	1667	$\chi^2 = 19.2$

Table 7.6. Observed and expected counts of racial group members (2000 GSS data)

population proportions, we may conclude that the data did not come from a simple random sample. In particular, based on the contribution to the chi square statistic made by the black subpopulation (15.63), it seems clear that blacks were undersampled. As this example demonstrates, the lack-of-fit test can therefore be used, when population proportions are known for a particular variable, to determine whether the sample is a simple random subset of the population.

7.3 Conclusions

In this chapter, we developed two statistical tests that can be applied to nominal and ordinal variables. The chi square test of independence is used to determine whether two variables are associated with one another or are independent. Given that the variables being considered may be nominal, the test is designed to determine whether any sort of patterning is present in the data. Thus, the test is often also used with ordinal data to capture nonlinear patterning.

The lack-of-fit test can be used to determine whether the proportions of sample members in each category of a nominal (or other) variable are consistent with a set of known or hypothesized proportions of population members in those categories. A discrepancy can be used to show either (1) that a set of hypothesized proportions is incorrect, or (2) that a sample is not representative of a population (i.e., it is not a simple random sample).

Although these chi square tests are extremely useful, especially when the level of measurement of variables is nominal, or when one is interested in determining whether nonlinear relationships may exist between variables, chi square tests do suffer from two important limitations. First, chi square test statistics are somewhat unstable when cell counts are small. A common rule of thumb is not to trust tests when any cells are expected to have fewer than five observations. Second, chi square tests are highly sensitive to sample sizes. Consider the calculation of the chi square: in the numerator of the calculation of each cell's contribution to the chi square statistic, the difference between the observed and expected counts is squared, but the denominator uses a single number. Thus, as the sample size increases, the numerator tends to grow at a faster rate than the denominator, making it easier to obtain large chi square statistics in large samples.

7.4 Items for Review

- Independence of events and the implications for crosstab cells
- Expected cell counts
- Chi-square test of independence
- Chi-square lack-of-fit test
- Problems with chi square tests

7.5 Homework

1. Reconstruct Table 7.3 to include column percentages rather than row percentages. Interpret.
2. I rolled a single six-sided die 1,000 times and obtained the following counts for each number 1–6 (respectively): 152, 163, 146, 173, 171, 195. Is the die weighted?
3. Is political party affiliation associated with happiness? The following data are from the 2006 GSS:

	Happiness			
Party	Unhappy	Somewhat Happy	Very Happy	Total
Democratic	69	304	119	492
Independent	67	264	128	459
Republican	41	239	189	469
Total	177	807	436	1420

4. Are health and happiness related? The following data are from the 2006 GSS:

	Happiness			
Health	Unhappy	Somewhat Happy	Very Happy	Total
Poor	23	39	10	72
Fair	59	160	52	271
Good	72	418	185	675
Excellent	23	190	189	402
Total	177	807	436	1420

5. 51 % of the population is female, and 49 % is male. In the GSS data, there are 12,038 males and 14,190 females. Is the sample consistent with these population proportions?
6. Of a sample of 50 men and 40 women, 60 % of the men voted, while 45 % of the women voted. Are sex and voting propensities independent?

7. Is marital status associated with political party affiliation?

	Party			
Marital Status	Democrat	Independent	Republican	Total
Married	192	204	289	685
Widowed	45	23	42	110
Div./Sep.	105	92	78	275
Never Married	150	140	60	350
Total	492	459	469	1420

8. In the previous problem, incorporate row or column percentages as appropriate and interpret.
9. Some argue that certain political positions are aligned within political parties. For example, Republicans tend to be both anti-abortion and pro-death penalty, while Democrats tend to be pro-choice and anti-death penalty. If this alignment is true, we would expect to see an association between how people respond to questions about these positions. Below is the data from the GSS for 2010. Based on the data, what would you conclude about the hypothesis about positions?

	Death Penalty			
Abortion	Pro	Anti	Don't Know	Total
Pro-Choice	358	156	21	535
Anti Abortion	449	203	31	683
Don't Know	17	9	7	33
Total	824	368	59	1251

10. Are births distributed equally across months, or is there some patterning? The following table shows the number of individuals born in each month in the GSS:

Month	Births
January	2651
February	2535
March	2709
April	2474
May	2482
June	2650
July	2694
August	2856
September	2781
October	2662
November	2591
December	2610
Total	31,695

Chapter 8
Comparing Means Across Multiple Groups: Analysis of Variance (ANOVA)

The independent samples t tests we discussed in Chap. 6 allow us to compare the means of two groups on some continuously-distributed outcome of interest (e.g., income, age, education, etc.). Often, however, we are interested in comparing the means of more than two groups. For example, we may be interested in comparing mean educational attainment across racial groups, with race measured as "White," "Black," and "Other." In cases such as these, we *could* compare each pair of means using a series of t tests, but this would require the computation of $\binom{g}{2}$ tests. This number becomes unreasonably large quickly as the number of groups, g, increases. For example, with 5 groups, we would have to perform $\binom{5}{2} = 10$ tests. Furthermore, it may be the case that you are uninterested in pairwise differences: Your theory may specifically ask simply whether there are racial differences in educational attainment and not specify which races differ from others. Thus, we need a test that simultaneously compares all means and tells us whether there is variation in the means across a number of groups. Analysis of Variance (ANOVA) is one approach to answering this type of question.

8.1 The Logic of ANOVA

The purpose of ANOVA is to determine whether differences *between* group means are large after accounting for differences in the variances *within* groups that may lead to highly variable group means from sample to sample. ANOVA is called "Analysis of Variance," because it accomplishes the comparison of differences between group means by decomposing the total variance in the outcome measure into within-group variance and between-group variance. If the between-group variance is sufficiently larger than the within-group variance, then the F test that stems from ANOVA calculations leads to the conclusion that there are differences between the means of the groups in the population.

As a demonstration of *how* ANOVA works, suppose we had data on weekly study hours for college students by class (excluding freshmen), and the data looked like

S.M. Lynch, *Using Statistics in Social Research: A Concise Approach*,
DOI 10.1007/978-1-4614-8573-5_8, © Springer Science+Business Media New York 2013

Sophomore	Junior	Senior
4	6	8
4	6	8
4	6	8
4	6	8
4	6	8
4	6	8

Table 8.1. Hypothetical distribution of study hours by year in college

Sophomore	Junior	Senior
4	4	4
4	4	4
6	6	6
6	6	6
8	8	8
8	8	8

Table 8.2. Another hypothetical distribution of study hours by year in college

those presented in Table 8.1. The table shows the study hours for 18 students, six from each class (sophomore year through senior year). In this example, it is clear that there is a difference in studying habits by college year. Why? Because *every* sophomore spent 4 h studying, *every* junior spent 6 h studying, and *every* senior spent 8 h studying per day.

Consider, in contrast, the data presented in Table 8.2. In the data presented in that table, it seems that study hours do not depend on college class at all. Instead, of 6 students in each class, 2 spend 4 h per week studying, 2 spend 6 h per week studying, and 2 spend 8 h.

In both of these examples, the overall sample mean is the same:

$$\frac{4+4+4+4+6+6+6+6+8+8+8+8}{18} = 6 \tag{8.1}$$

Furthermore, in both examples, the overall sample variance is the same:

$$\frac{\sum_{i=1}^{n}(x_i - \bar{x})^2}{n-1} = 2.82. \tag{8.2}$$

However, the two samples clearly exhibit different patterns. What differs between the two samples is the composition of study hours within versus between year of study. In the first example, it is clear that knowing class year perfectly predicts study hours, but in the second example, knowing a student's class tells us nothing about his/her study habits.

ANOVA helps us formally differentiate these two samples by decomposing the total sample variance into two components—the variance within groups versus the variance between groups—and then assessing their relative contributions to the total sample variance.

8.2 Some Basic ANOVA Computations

In order to assess the relative contributions of between vs. within group variance to the total sample variance, we must compute both types of variance, as well as the total sample variance. In Chap. 4, we discussed how to compute total sample variance: the computation involves (1) computing the overall sample mean, and (2) computing the average squared deviation of each individual value from the overall sample mean.

To decompose this total sample variance into between-group and within-group variance, we must consider how much individuals within groups deviate from their own group means and how much the group means differ from each other. In order to make this comparison, let's exclude the denominator of the variance calculations for a moment and just obtain the numerators of the variance calculations. These calculations are called the "sums of squares," and there are three such calculations. The first is the "total sum of squares" (abbreviated SST). This sum of squares is the numerator of the sample variance and is calculated as:

$$SST = \sum_{i=1}^{n} \left(x_i - \bar{\bar{x}}\right)^2 \tag{8.3}$$

$$= \sum_{g=1}^{G} \sum_{i=1}^{n_g} (x_{ig} - \bar{\bar{x}})^2 \tag{8.4}$$

All of the terms in the first line should be familiar, except $\bar{\bar{x}}$. In ANOVA $\bar{\bar{x}}$ replaces $\bar{x}$ as the overall sample mean. The double-bar differentiates the "grand mean" from the group means that are used in the other sums of squares calculations.

The second line in the SST calculation is an alternative way to write this sum of squares. In this expression, G refers to the total number of groups, n_g is the number of persons in group g. x_{ig} is therefore the value of x for the ith person in group g. This expression explicitly shows that the total sum of squares involves summing the squared deviations of each member (i) of each group (g) from the grand mean.

The second sum of squares calculation is the within-group sum of squares (abbreviated SSW). It measures the amount of variation that exists within each group and is calculated as:

$$SSW = \sum_{g=1}^{G} \sum_{i=1}^{n_g} \left(x_{ig} - \bar{x}_g\right)^2. \tag{8.5}$$

Sophomore	Junior	Senior
4	6	8
4	6	8
4	6	8
4	6	8
4	6	8
4	6	8
$\bar{x}_1 = 4$	$\bar{x}_2 = 6$	$\bar{x}_3 = 8$
$\sum_{i=1}^{6}(x_{i1} - \bar{x}_1)^2 = 0$	$\sum_{i=1}^{6}(x_{i2} - \bar{x}_2)^2 = 0$	$\sum_{i=1}^{6}(x_{i3} - \bar{x}_3)^2 = 0$
	$\downarrow$	
	$SSW = 0 + 0 + 0$	

Table 8.3. Sums of squares for the first hypothetical set of data. Note: $\bar{\bar{x}} = 6$; $SST = \sum_{g=1}^{G=3} \sum_{i=1}^{n_g}(x_{ig} - \bar{\bar{x}})^2 = 48$.

This equation looks similar to the second version of the SST equation, however, $\bar{\bar{x}}$ has been replaced with $\bar{x}_g$. This is the mean of x for group g. Thus, the key difference between SSW and SST is that each individual's value of x is deviated from its *group* mean rather than the grand mean.

The third sum of squares calculation is the between-group sum of squares (abbreviated SSB). It measures the extent of variation between the group means and is calculated as:

$$\sum_{g=1}^{G} n_g(\bar{x}_g - \bar{\bar{x}})^2. \tag{8.6}$$

The calculation simply involves computing the squared deviation of each group's mean ($\bar{x}_g$) from the grand mean ($\bar{\bar{x}}$), weighting this quantity by the group size (n_g), and summing across groups.

To make the idea of sums of squares concrete, consider again the data presented in Table 8.1. Table 8.3 shows the original data, as well as the group means, the overall sample mean, and the sums of squares calculations. From this table, we can see that the sums of squares within each group is 0 (for all groups), while the total sums of squares is 48. In other words, while there is variation across the sample, there is no variation of study hours within classes.

It can be shown that:

$$SST = SSB + SSW. \tag{8.7}$$

In these hypothetical data, then, given that SSW is 0 and SST is 48, SSB must be 48.

Sophomore	Junior	Senior
4	4	4
4	4	4
6	6	6
6	6	6
8	8	8
8	8	8
$\bar{x}_1 = 6$	$\bar{x}_2 = 6$	$\bar{x}_3 = 6$
$\sum_{i=1}^{6}(x_{i1} - \bar{x}_1)^2 = 16$	$\sum_{i=1}^{6}(x_{i2} - \bar{x}_2)^2 = 16$	$\sum_{i=1}^{6}(x_{i3} - \bar{x}_3)^2 = 16$
	$\downarrow$	
	$SSW = 16 + 16 + 16$	

Table 8.4. Sums of squares for the second hypothetical set of data. Note: $\bar{\bar{x}} = 6$; $SST = \sum_{g=1}^{G=3} \sum_{i=1}^{n_g}(x_{ig} - \bar{\bar{x}})^2 = 48$.

In contrast, what happens if we apply these calculations to the second set of hypothetical data? Table 8.4 shows these results. In this case, the SSW is 48, the SSB is 0, and the SST is (still) 48.

Comparing these two sets of results is revealing. In the first case, SSB = SST; in that second case, SSW = SST. In other words, in the first scenario, all of the variation in study hours in the overall sample was attributable to between group variation. All of the sophomores looked the same, all of the juniors looked the same, and all of the seniors looked the same, but the classes differed from each other. The only variation across the sample was attributable to year of study. In the latter case, on the other hand, there was considerable variation in study habits between members of the same class, but, overall, the classes looked identical.

These results suggest a simple statistic to differentiate these samples, namely, the proportion of the total sample variance (or sums of squares) that is between-group variance. This statistic is called R^2 and is computed as:

$$R^2 = \frac{SSB}{SST} \equiv 1 - \frac{SSW}{SST}. \tag{8.8}$$

In the first example, $R^2 = 1$; in the latter, $R^2 = 0$. In other words, 100 % of the variance in the first sample was explained by the groupings we chose (year of college), but 0 % of the variance in the second sample was explained by our groupings. R^2 is an important quantity in statistics, and we will see it again when we discuss the correlation coefficient and regression modeling. However, the R^2 statistic we have derived leaves several things to be desired. Importantly, R^2 by itself tells us nothing about whether the observed between-group differences we've observed could be attributable to sampling fluctuation. Second, how high of an R^2 do we need before we can conclude that, in the population, the outcome of interest (differences in sample means) in fact varies across the groups we're interested in differentiating? The F statistic from an ANOVA table can help us with this determination.

8.3 A Real ANOVA Example

In the examples presented in the previous section, either all of the variation in the sample was explained by college class or none of it was. In reality, we can never fully explain variation in an outcome on the basis of a unidimensional classification of people. Instead, real data tend to produce R^2 values that are substantially less than 1: some of the variance in the sample is due to between group differences, while some is due to within group variance. An ANOVA table is constructed in order to show these relative contributions to the overall sample variance and to present an F statistic, which can be assessed to determine whether the ratio of between group variance to within group variance is sufficiently large to reject a null hypothesis that the means of all groups are equal.

A generic ANOVA table follows the format shown in Table 8.5. As the table shows, if one knows (1) the sample size, (2) the number of groups, (3) the total sums of squares, and (4) one other of the two types of sums of squares, the remainder of the table can be computed with no additional information. The first column of the table simply labels the source of the sums of squares. The second column presents the three sums of squares discussed previously. The third column presents the degrees of freedom associated with the sums of squares. Notice that the degrees of freedom for the between and within sums of squares sum to the total degrees of freedom, which is simply the denominator of the total sample variance formula. The fourth column presents the mean sums of squares, which are simply the various sums of squares divided by their degrees of freedom. Finally, the fifth column presents the F statistic, which is simply the ratio of the between and within mean squares.

Source	Sum of Squares	DF	Mean Squares	F
Between	$\sum_{g=1}^{G} n_g \times (\bar{x}_g - \bar{\bar{x}})^2$	$G-1$	$MSB = \frac{SSB}{df(B)}$	$\frac{MSB}{MSE}$
Within	$\sum_{g=1}^{G} \sum_{i=1}^{n_g} (x_{ig} - \bar{x}_g)^2$	$n-G$	$MSE = \frac{SSW}{df(W)}$ *	
Total	$\sum_{g=1}^{G} \sum_{i=1}^{n_g} (x_{ig} - \bar{\bar{x}})^2$	$n-1$	(Sample Variance**)	

* The notation for the within sum of squares divided by its degrees of freedom changes: MSE is called the mean squared error—it is the portion of the variance unexplained by the grouping variable.

** Generally not included in the actual table, but is implicit.

Table 8.5. Format for a generic one-way ANOVA table.

Source	Sum of Squares	DF	Mean Squares	F
Between	483.91	2	241.95	26.57 ($p < .001$)
Within	12903.73	1417	9.11	
Total	13387.64	1419	9.43	$R^2 = .036$

Table 8.6. ANOVA example: Race differences in educational attainment (2006 GSS data).

The F statistic is a measure of the extent to which the total variance in the variable of interest is accounted for by the grouping variable. The null hypothesis in the F test is that the group means are equal; therefore, none of the variance is from group differences. In repeated sampling, of course, differences between group means will sometimes be evident, just as we have discussed throughout earlier chapters. Thus, the F statistic can help us assess, under the assumption that there are no differences in group means, how likely it is that we could obtain the difference in means we observed in a random sample. As with other statistical tests, the rarer our observed sample would be under this null hypothesis, the more confident we can be in rejecting it.

The F statistic can be assessed using an F table, just as you have assessed z scores using a z (normal distribution) table, t scores using a t table, and χ^2 statistics using a χ^2 table—the F distribution is simply another distribution. Like the t and χ^2 distributions, the F distribution has degrees of freedom associated with it; however, the F distribution has a pair of degrees of freedom, generally called "numerator" and "denominator" degrees of freedom. These degrees of freedom for ANOVA are $df(B)$ and $df(W)$ respectively.

Table 8.6 an ANOVA table for education by race obtained from Stata using the 2006 GSS data. Race is measured as a three category variable (black, white, other), and educational attainment is measured in years. Mean attainment for blacks was 12.66 (s.d. = 3.15); the mean for whites was 13.72 (s.d. = 2.79); the mean for others was 12.14 (s.d. = 4.03). The F test in the ANOVA table tests the hypothesis that there is no difference in these means. As the results show, we can reject this null hypothesis, with the implication being that there is at least one mean that is different from the others. Realize, however, the test does not tell us which mean(s) differ(s) from which.

8.4 Conclusions

In this chapter, we have developed a statistical approach to examining differences in means between multiple groups. The method—ANOVA—extends the independent samples t test to handle more than two groups simultaneously. The null hypothesis of interest in ANOVA is that all group means are equal. Rejecting this null hypothesis tells us only that at least one mean differs from the other; it does not tell us which mean or means differ from which.

ANOVA modeling can be extended to simultaneously examine more than one grouping variable (called "two-way" ANOVA). We did not discuss such approaches; instead, we will discuss regression modeling approaches to incorporating additional variables. Regression modeling and ANOVA modeling are similar in many ways, but regression modeling is generally more flexible and more commonly used across the social sciences.

8.5 Items for Review

- Between, within, and total sums of squares
- R^2 (Explained variance)
- The F test and F distribution
- Numerator and denominator degrees of freedom
- The ANOVA table

8.6 Homework

1. Below is a partially-completed ANOVA table examining racial differences in family income from the 2006 GSS. Complete the table, state the null hypothesis, and draw a conclusion.

Source	Sum of Squares	DF	Mean Squares	F
Between	78169.41	2		
Within				
Total	2328739.13	1419		$R^2 =$

2. Below are data from a hypothetical clinical trial testing the effectiveness of a new drink, the manufacturers of which claim the drink boosts energy levels. The clinical trial involved three groups: a treatment group (the members of which received the beverage), a placebo group (the members of which received a "fake" beverage), and a control group (the members of which received no beverage at all). Assume that participants in the trial were randomly selected

from the population and were randomly assigned to treatment/placebo/control groups. Also assume that participants' energy levels were measured before and after receiving the beverage, and the reported values in the table below represent the increase (or decrease) in energy level after receiving the drink (or, in the case of the control group, after NOT receiving anything). Is there evidence that the energy drink works?

Person	Treatment Group		
	Treatment	Placebo	Control
1	0	1	2
2	3	1	0
3	1	2	0
4	5	3	3
5	2	0	1

3. Below is a table with descriptive statistics for health by level of education, where education is defined by three groups: those with less than a high school diploma, those with a high school diploma, and those with education beyond high school. Using these data, construct the appropriate ANOVA table and test the null hypothesis that mean health is constant across levels of education.

Education	n	Mean Health	s.d. for Health
Less than H.S.	223	1.60	.88
High School	372	1.93	.79
More than H.S.	825	2.12	.78
All	1420	1.99	.82

4. Below is a partially-completed ANOVA table examining political party (Democrat, Independent, Republican) differences in happiness (assume happiness is continuous). Complete the table, state the null hypothesis, and draw a conclusion.

Source	Sum of Squares	DF	Mean Squares	F
Between		2		
Within	553.11			
Total	565.76	1419		$R^2 =$

5. Below is a table with descriptive statistics for life satisfaction by first term of three US Presidents: Reagan (1981–1984), Bush Sr. (1989–1992); and Clinton (1993–1996). Is there variation in mean life satisfaction by presidency? (Hint: you will have to determine the overall mean for life satisfaction using the subsample means).

President	n	Mean satisfaction	s.d. for satisfaction
Reagan	2356	22.73	4.61
Bush Sr.	1085	23.06	4.70
Clinton	580	22.80	4.69

6. Are their regional differences in health, based on the region in which a respondent was raised? Below are data on health (0 = poor…3 = excellent) by region at age 16 from the 2010 GSS (persons age 40).

Person	Region at 16			
	Northeast	Midwest	South	West
1	2	1	2	1
2	2	1	2	2
3	3	2	2	3
4	3	2	2	3
5		3	2	
6			2	
7			2	
8			3	

Chapter 9
Correlation and Simple Regression

The methods we have discussed so far allow us to examine whether means of a numeric variable (interval/ratio) vary across two or more groups measured at the nominal level and whether two or more nominal, or possibly ordinal variables, are associated. However, we often want to determine whether two continuous variables are related. In these cases, none of the methods we have covered is appropriate. Instead, we may turn to two alternate methods: the Pearson correlation coefficient and the simple linear regression model. These methods form the basis for the more widely used multiple regression model, which we will discuss in the next chapter.

9.1 Measuring Linear Association

9.1.1 The Covariance

Consider the hypothetical variables x and y shown in the scatterplots in Fig. 9.1. In the upper left plot in the figure, x and y are clearly very strongly related to each other: larger values of x correspond to larger values of y (and vice versa). In the upper right plot, the two variables do not appear to be patterned. In the lower left plot, x and y are very strongly related, but in the opposite direction compared to the upper left plot. Here, larger values of x correspond with smaller values of y. Finally, in the lower right plot x and y appear to be positively related, but not as strongly as in the first plot.

How can we measure the extent of the linear association between two variables like those in each plot? The covariance is a first measure of the strength of linear association between two continuous variables. It is computed as:

$$cov(x, y) = \frac{\sum_{i=1}^{n}(x_i - \bar{x})(y_i - \bar{y})}{n-1}. \tag{9.1}$$

S.M. Lynch, *Using Statistics in Social Research: A Concise Approach*,
DOI 10.1007/978-1-4614-8573-5_9, © Springer Science+Business Media New York 2013

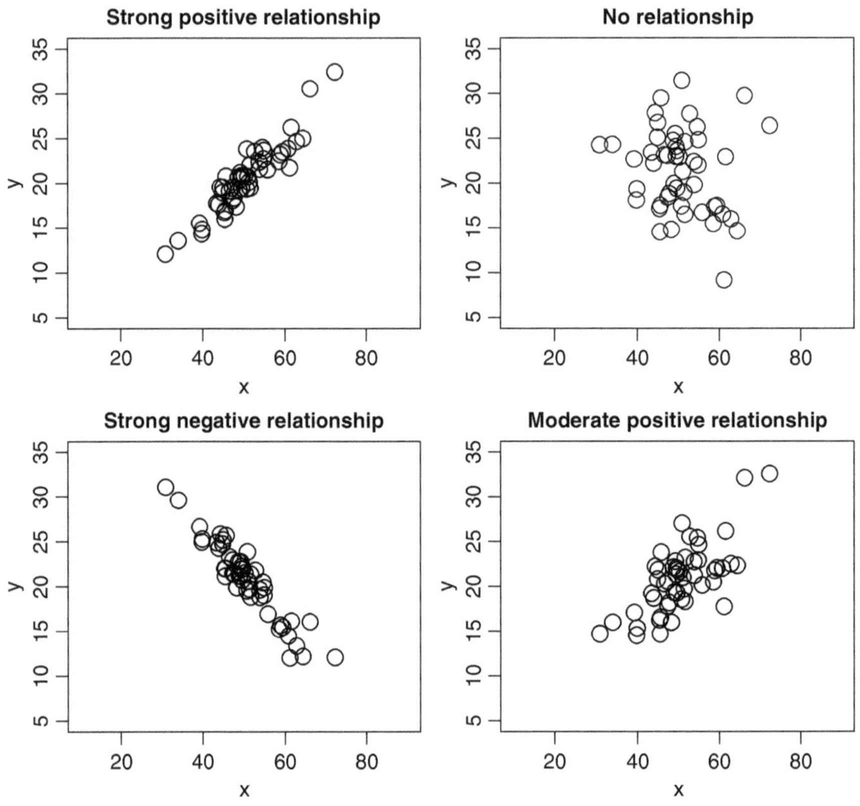

Fig. 9.1. Scatterplot of hypothetical variables x and y.

This equation looks like the variance formula from Chap. 4, but it contains information on two variables (x and y), not just one. Specifically, whereas the numerator of the variance formula involves summing the squared deviations of the sample values of a variable from its mean, the numerator of the covariance requires summing the product of x deviations from its mean with y deviations from its mean. In both the variance and covariance calculations, the numerator is then "averaged" by dividing by $n - 1$.

If two variables, x and y, are linearly related, then the covariance will be large, either positively or negatively. Why? Consider the relationship between x and y again with some of the components of the covariance calculation superimposed in the plot (see Fig. 9.2). If x and y are linearly associated, then whenever x is far from its mean (i.e., $x - \bar{x}$ is large), y should also be far from its mean (i.e., $y - \bar{y}$ is large). The product of two large numbers is large itself, and the sum of a set of large numbers will be large, so the covariance will be large. If x and y are not linearly related, then when $x - \bar{x}$ is large, $y - \bar{y}$ will, on average, be close to 0, and when $y - \bar{y}$ is large, $x - \bar{x}$ will, on average, be close to 0. The product of large numbers with numbers close to 0 is small, so is their sum, and so the covariance will be small.

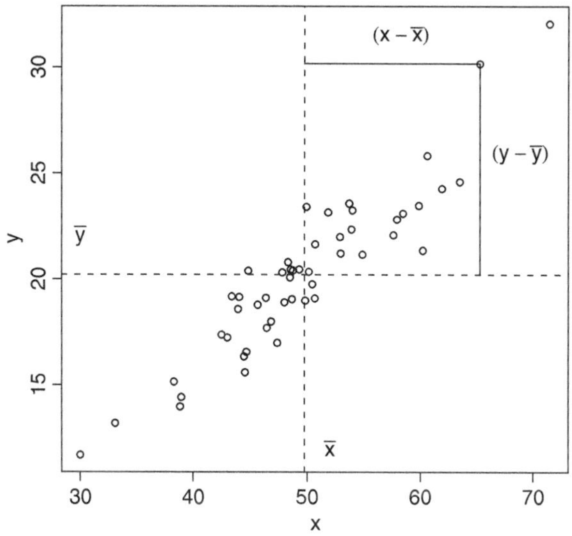

Fig. 9.2. Scatterplot of hypothetical variables x and y with deviations from means superimposed for a single observation.

9.1.2 The Pearson Correlation

A key limitation of using the covariance to assess the strength of the relationship between two variables is that its scale is a function of the scale of the variables used to compute it. If you change the scale, the size of the covariance will change. The correlation (denoted r) corrects for this scale problem by dividing the covariance by the standard deviation of each variable:

$$r = \frac{\sum_{i=1}^{n}(x_i - \bar{x})(y_i - \bar{y})/(n-1)}{sd(x)sd(y)}. \tag{9.2}$$

Performing this division constrains the correlation to be between -1 and $+1$. Thus, the correlation represents the expected standard unit change in one variable for a standard unit change in the other variable. In this metric, a perfect positive association is represented by a correlation of +1, a perfect negative association is represented by a correlation of -1, and a 0 represents no linear association at all.

Figure 9.3 is a scatterplot of education and family income from the 2004 GSS (unmarried persons and persons with 0 income have been excluded). Education and income appear to be related linearly, but the scale of the variables makes it difficult to discern the strength of the relationship.

Figure 9.4 shows the same information, but after the two variables have been standardized. The figure contains a 45-degree line; if the data were to fall along this line, the correlation between education and income would be 1. Instead, the

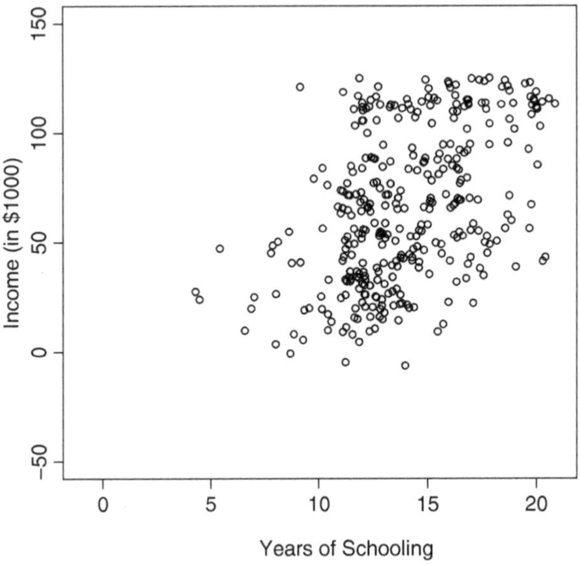

Fig. 9.3. Scatterplot of education and income (2004 GSS Data).

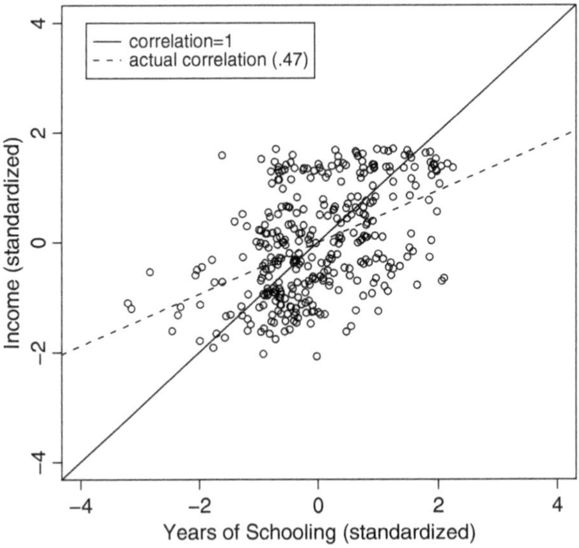

Fig. 9.4. Scatterplot of standardized education and income with reference lines

correlation between education and income is .47, which is considered to be a moderately strong linear relationship between the two variables. In general, in social science research a correlation between .1 and .3 is considered a weak association, a correlation between .3 and .6 is considered to be a moderate association, and a correlation above .6 is considered to be a strong association.

9.1.3 *Confidence Intervals for the Correlation*

In previous chapters, we have seen that sampling variability produces sample means ($\bar{x}$) that differ from the true population mean (μ). Sampling variability similarly produces estimates of the correlation (r) that differ from the true value in the population (ρ). In previous chapters, we constructed confidence intervals around our sample estimate in order to make inference about a reasonable value for μ. We can construct a similar interval estimate for the correlation.

However, the correlation is bounded by -1 and 1, and even though its sampling distribution becomes more normal in appearance as sample sizes increase, its distribution will remain bounded and slightly skewed so long as the correlation is not exactly 0. Ronald Fisher, a prolific and influential statistician from the early twentieth century, proposed a transformation of the correlation that has an approximately normal distribution and allows for construction of reasonable confidence intervals. The process of interval construction involves the following steps:

1. Transform r into z_f using: $z_f = .5\,[\ln(1+r) - \ln(1-r)]$ (z_f is "Fisher's z")
2. Compute the standard error of z_f: $\hat{\sigma}_{z_f} = \frac{1}{\sqrt{n-3}}$
3. Compute the interval in the z_f metric as:

$$z_f \pm z_{\alpha/2}\hat{\sigma}_{z_f}$$

4. Convert the lower and upper bounds (b) back to the correlation scale using:

$$b = \frac{e^{2z} - 1}{e^{2z} + 1} \tag{9.3}$$

How does this transformation work to create a more normally distributed sampling distribution for the correlation? Consider the calculation. When r is 0, the transformation is:

$$z_f = .5\,[\ln(1+r) - \ln(1-r)] \tag{9.4}$$

$$= \frac{1}{2}\ln\left(\frac{1+r}{1-r}\right) \tag{9.5}$$

$$(1/2)\ln(1) \tag{9.6}$$

$$= 0 \tag{9.7}$$

When r is 1, the transformation is:

$$z_f = \frac{1}{2}\ln\left(\frac{1+r}{1-r}\right) \tag{9.8}$$

$$= (1/2)\ln(2/0) \tag{9.9}$$

$$\approx \infty \tag{9.10}$$

When r is -1, the transformation is:

$$z_f = \frac{1}{2}\ln\left(\frac{1+r}{1-r}\right) \tag{9.11}$$

$$= (1/2)\ln(0/2) \tag{9.12}$$

$$\approx -\infty \tag{9.13}$$

These latter two sets of equations are approximate, because both division by 0 and the log of 0 are undefined. However, as r *approaches* these values, z_f approaches positive or negative infinity. In other words, Fisher's z transformation stretches the correlation's bounds from ± 1 over the real line.

In the education and income data shown in the figures, the correlation was .47, and the sample size was 342. Fisher's z, then, is:

$$z_f = .5\,[\ln(1+r) - \ln(1-r)] \tag{9.14}$$

$$= .5\,[\ln(1.47) - \ln(.53)] \tag{9.15}$$

$$= .5(.385 - (-.635)) \tag{9.16}$$

$$= .51 \tag{9.17}$$

The standard error is $(1/\sqrt{342-3}) = .0543$. A 95 % confidence interval for z_f can be computed by adding and subtracting 1.96 standard errors from the estimate z_f:

$$\begin{gathered}[.51 - (1.96)(.0543)\ ,\ .51 + (1.96)(.0543)] \\ [.404\ ,\ .616]\end{gathered} \tag{9.18}$$

We can now transform these bounds back into the correlation scale to obtain:

$$b_L = \frac{e^{2z} - 1}{e^{2z} + 1} \tag{9.19}$$

$$= \frac{e^{2(.404)} - 1}{e^{2(.404)} + 1} \tag{9.20}$$

$$= .383 \tag{9.21}$$

and

$$b_U = \frac{e^{2z} - 1}{e^{2z} + 1} \tag{9.22}$$

$$= \frac{e^{2(.616)} - 1}{e^{2(.616)} + 1} \tag{9.23}$$

$$= .548 \tag{9.24}$$

So, our 95 % confidence interval is [.383, .548]. Our interval estimate does not contain 0, and therefore we may conclude that a linear relationship between education and income exists in the population (the population correlation is not 0).

9.1.4 *Hypothesis Testing on the Correlation*

We can develop a hypothesis test for the correlation in the Fisher's z metric as an extension of the original z test formula presented in Chap. 6:

$$z = \frac{z_f - H_0}{\hat{\sigma}_{z_f}}. \tag{9.25}$$

The usual null hypothesis is that the correlation $\rho = 0$ in the population; that is, there is no linear association between the two variables. In the example from the previous section, Fisher's $z_f = .51$ and $\hat{\sigma}_{z_f} = .0543$. Furthermore, when $\rho = 0$, $z_f = 0$. Thus:

$$z = \frac{.51 - 0}{.0543} \tag{9.26}$$

$$= 9.39. \tag{9.27}$$

$$\tag{9.28}$$

This z score is more than large enough to reject the null hypothesis that $\rho = 0$ in the population. Thus, we conclude that there is a linear association between education and income in the population.

9.1.5 *Limitations of the Correlation*

The correlation is useful as a first step in examining the relationships between variables in an analysis. However, it is limited in several respects. First, without modifications, the correlation does not tell us the extent to which a relationship between two variables may be contingent on a third variable. That is, what part of the correlation—if not all of it—is spurious (i.e., false)? For example, there is a strong correlation between ice cream consumption rates and violent crime rates, but this relationship is entirely explained by temperature.

Second, the correlation only measures linear association, and so it does not tell us anything about the relationship between variables that are not linearly related.

Third, the correlation is agnostic with respect to the direction of the causal relationship between variables. Of course, it is well-known and often repeated that "correlation does not prove causation," and we will discuss this in the next chapter in some depth. Yet it is often the case that we may wish to single out one variable as a "cause" and another as an "effect" for theoretical reasons. In such cases, the correlation is not theoretically pleasing.

To be sure, no method can directly address all of these limitations, especially the last. However, linear regression analysis comes closer to addressing these issues—at least theoretically—and answering the types of questions we are interested in social science.

9.2 Simple Linear Regression

Linear regression is probably the most important method of quantitative data analysis you will learn. It is one of the most widely used methods in scientific research and is the foundation of more complex methods. The logic of regression analysis is simple; the mathematics can range from relatively simple to relatively complex. For the purposes of this book, I am primarily interested in introducing the purpose of regression modeling and providing a basis for understanding how regression modeling works and what kind of information it yields. Thus, I will keep the mathematics as simple as possible.

9.2.1 Logic of Simple Regression

The basic logic of regression analysis is as follows. Suppose I have some variable y that I think is affected by another variable x, and I am interested in (1) determining whether there is a relationship between x and y and (2) quantifying its strength (note that I have explicitly stated that x causes y and not the other way around). If I believe the relationship between x and y to be linear, I could specify the relationship as:

$$y = \alpha + \beta x, \tag{9.29}$$

where α is the "intercept," and β is the "slope." Notice that this equation looks very similar to the slope-intercept form of the equation for a line from algebra: $y = mx + b$. The only differences are that b has been replaced by α, m has been replaced by β, and the two terms on the righthand side have traded places. These are only cosmetic differences: the idea behind both equations is the same. There is some value of y for the case in which $x = 0$ (α; the intercept), and there is a relationship between x and y such that a one-unit increase in x translates into—produces—a β unit increase in y. To make this idea concrete, consider the relationship between education and income. We may think that obtaining more education increases income. Thus, x is education, and y is income, and we expect that β is positive. Since income is generally positive, even at minimal levels of schooling, we might expect that α is also positive.

There are a few key problems with the model specified above. First, this specification is deterministic, meaning that a change in x produces a guaranteed β unit change in y. This is almost never the case in the real world. Second, and similarly, this model suggests that x is the *only* factor that is important in affecting y. This is also almost never the case in the real world. For example, education may be *a* cause of income, but so is winning the lottery and receiving an inheritance from a dead relative. Smoking may be a cause of lung cancer, but so is radon and exposure to asbestos. Lack of exercise may be a cause of obesity, but so is overeating and genetics. We certainly do not expect to have a single cause for any outcome. Otherwise, the scatterplot in Fig. 9.3 would have shown exactly one value of income for each value of education.

These limitations suggest that, to make the model more realistic, we should make the model a little more flexible, and so, we can modify it a little:

$$y_i = \alpha + \beta x_i + e_i. \tag{9.30}$$

In this specification, we have added the subscript i and an "error" term e_i. This model now says that the value of y for any individual (i) is a function of an intercept (which is constant for everyone), an effect of x (β; also constant for everyone), and some individual-specific error. In other words, under this specification we do *not* expect that x is the only cause of y nor do we expect that a one-unit change in x is necessarily associated with a β unit increase in y for every individual—we *do* expect a β unit increase in y for a one unit increase in x, but this is only an *average* effect and may not hold true for everyone.

For example, consider again the relationship between education and income. If we expect that income is affected by education, such that persons with more education make more income than persons with less education, we may wish to estimate the model:

$$\text{Income}_i = \alpha + \beta \times \text{Education}_i + e_i. \tag{9.31}$$

If we estimate this model using a subset of the GSS data, we obtain:

$$\text{Income}_i = -11.8 + 5.6 \times \text{Education}_i + e_i. \tag{9.32}$$

This result implies that each person's income can be recovered from his education and his unique error term along with the estimated intercept and slope. For persons not in the sample, we may estimate (predict) their income from their education using the predicted value:

$$\widehat{\text{Income}} = -11.8 + 5.6 \times \text{Education}.$$

Notice in this equation, the error term has dropped out; the reason is that, for the linear regression model, while individuals are expected to have some error, the average error is expected to be 0. We are generally not terribly interested in the particular sample members we have, but instead we are interested in the general pattern. Here, the results say that each additional year of schooling is worth (on average) $5,600 more in income (recall that income is in $1,000 units).

Technically, the results show that mean income at each level of education is about $5,600 higher than the level of education below it. Given that we have observed *different* people at each level of schooling, we cannot truly conclude that increasing a person's education would necessarily lead to a higher income for him/her.

9.2.2 Estimation of the Model Parameters

How are the values of α and β estimated? In general, we would like estimates of the parameters that yield a prediction line that closely represents the pattern observed in a scatterplot of the data. Reconsider, for example, the two lines shown in Fig. 9.4. The 45-degree line (solid line) in the figure does not seem to do as good of a job at "fitting" the observed pattern as the dashed line. But, what makes the dashed line better?

There are many criteria that could be used to find estimates of α and β, but the most commonly used criterion is the "least squares" criterion. The least squares criterion is that we want values for the intercept and slope that minimize the sum of the squared error terms (the e_i) around the line implied by the parameter estimates:

$$G = min\left(\sum_{i=1}^{n}(y_i - \hat{y}_i)^2\right), \tag{9.33}$$

where y_i is the observed value of y for person i, and $\hat{y}_i$ is the model-predicted value of y for person i. Given that $\hat{y}_i = \alpha + \beta x_i$, we can substitute this linear combination in for $\hat{y}_i$:

$$G = min\left(\sum_{i=1}^{n}(y_i - (\alpha + \beta x_i))^2\right), \tag{9.34}$$

Source	Sum of Squares	DF	Mean Squares	F
Model	$\sum_{i=1}^{n}(\hat{y}_i - \bar{y})^2$	1	$MSR = \frac{SSR}{df(R)}$	$\frac{MSR}{MSE}$
Error	$\sum_{i=1}^{n}(y_i - \hat{y}_i)^2$	$n-2$	$MSE = \frac{SSE}{df(E)}$	
Total	$\sum_{i=1}^{n}(y_i - \bar{y})^2$	$n-1$	(Sample Variance)	

Table 9.1. Generic ANOVA table for a simple regression model.

We can find the best values of α and β by using a little calculus. In particular, the minimum (or maximum) of a function is the point where the slope of the curve implied by the function is 0. Thus, the function G is minimized by first taking its partial derivative with respect to α and then with respect to β. Next, we set these partial derivatives equal to 0 and solve to find the value of the parameters that minimize the function. If we do this, we find:

$$\hat{\alpha} = \bar{y} - \hat{\beta}\bar{x} \tag{9.35}$$

$$\hat{\beta} = \frac{\sum_{i=1}^{n}(x_i - \bar{x})(y_i - \bar{y})}{\sum_{i=1}^{n}(x_i - \bar{x})^2}. \tag{9.36}$$

Notice that we must find β before computing α. Also notice that the solution for β can be rewritten as:

$$\hat{\beta} = \frac{cov(x, y)}{var(x)}, \tag{9.37}$$

because the numerator is simply $n - 1$ times the covariance of x with y, and the denominator is $n - 1$ times the variance of x. Finally, notice that the parameters have a caret (hat) above them, indicating that they are *estimates* of the population parameters α and β.

9.2.3 Model Evaluation and Hypothesis Tests

Once estimates of the intercept and slope are obtained, we are usually interested in evaluating how well the model fits the data and testing hypotheses regarding the relationship between x and y in the population. The parameter estimates can be used for these purposes. First, we can construct an ANOVA table, much as we did in the previous chapter, for the regression model results. The regression ANOVA table follows the format shown in Table 9.1.

Source	Sum of Squares	DF	Mean Squares	F
Model	89240.19	1	89240.19	95.02
Error	319315.48	340	939.16	
Total	408555.7	341	(1198.11)	

Table 9.2. ANOVA table for model of income regressed on education (data from continued example)

This table differs only slightly from the ANOVA table shown in the previous chapter. First, the names for the sources of the sums of squares differ: for regression, we have "model" (or "regression") and "error" rather than "between" and "within." Second, the calculations of the sums of squares are in terms of the data (y), the model predicted values ($\hat{y}$), and the sample mean ($\bar{y}$). It is still true that $SST = SSR + SSE$ (total sum of squares equals regression sum of squares plus error sum of squares). In the regression context, the total sum of squares is partitioned into the part explained by the regression line ($\hat{y} - \bar{y}$) and the part not explained by the regression line ($y - \hat{y}$). The remaining calculations are carried out much as before, but note the change in degrees of freedom: the model degrees of freedom for the simple regression model are 1 (two parameters minus 1); the error degrees of freedom are $n - 2$ (sample size minus number of estimated regression parameters).

Table 9.2 shows the ANOVA table from the education-income example. From the model sum of squares and the total sum of squares, we can compute R^2 as we did in ANOVA; the result is .22, meaning that 22 % of the variance in income is explained by education (conversely, 78 % remains to be explained—is error). This is a pretty strong result: in social science research, we are usually happy to have a double digit R^2.

The F statistic from this table is 95.02, which has a corresponding p value of approximately 0. This result tells us that we can reject the null hypothesis that there is no linear relationship between education and income in the population. In other words, education and income are linearly related in the population.

In the simple regression model, with one predictor (x) variable, the F statistic suffices as a test of the relationship between x and y. In multiple regression, however, the F test is an "omnibus" test that simply tells us whether at least one of the x variables is related to y, just as the F in ANOVA did not tell us which groups' means differed; it only told us that at least one mean differed from the others.

In order to conduct hypothesis tests on particular parameters so that we may determine whether particular x variables are related to y, we need standard errors of the parameter estimates. Just as we expect sample means to vary from the population mean and from sample to sample, we may expect that estimated regression parameters will also vary from sample to sample. Indeed, the Central Limit Theorem applies to regression coefficients much the same way as it applies to means: the distribution of sample regression coefficients is asymptotically normal with a mean equal to the true population regression coefficient (slope), and a standard deviation equal to a function of the mean squared error of the regression (MSE; see the ANOVA table) and the variance of x. Specifically:

$$s.e.(\hat{\alpha}) = \sqrt{\frac{\sigma_e^2 \sum x_i^2}{n \sum (x_i - \bar{x})^2}} \tag{9.38}$$

$$s.e.(\hat{\beta}) = \sqrt{\frac{\sigma_e^2}{\sum (x_i - \bar{x})^2}}, \tag{9.39}$$

where σ_e^2 is estimated using the MSE.

In the education-income example, the standard error for α ($\alpha = -11.8$) was 8.15 and the standard error for β ($\beta = 5.6$) was .57. In previous z and t tests, we established a hypothesized value for the population parameter (mean; μ) and evaluated how probable it would be for our sample mean to occur if the hypothesized value were true. We follow the same strategy in regression modeling, and we usually test the hypothesis that $\beta = 0$. If $\beta = 0$, then there is no linear relationship between x and y in the population; therefore, if we can reject this null hypothesis, the implication is that there is a linear relationship between the two variables in the population. We can test a similar hypothesis for the intercept, but because the intercept corresponds to the expected value of y when $x = 0$, and $x = 0$ may be an unreasonable value, we generally aren't terribly interested in testing whether the intercept is 0. Given that the MSE is an estimate of σ_e^2, the resulting test is a t test. In the education-income example, the t-test on the slope (β) is:

$$t = \frac{5.6 - 0}{.57} = 9.75. \tag{9.40}$$

The sample is large enough that we can use the z table to find the p value; that value is approximately 0, so we can reject the null hypothesis that $\beta = 0$ in the population. Education is linearly related to income.

9.3 Conclusions

In this chapter, we have developed statistical methods for assessing the relationship between two continuous variables, including the covariance, the correlation, and the simple regression model. As we discovered, the covariance is limited because of its measurement scale dependence. The correlation overcomes this limitation, but it suffers from its own limitations, including that it is agnostic with respect to the direction of causality between variables, and it cannot directly rule out spurious explanations for the association between variables. The simple regression model resolves, at least at a theoretical level, the first problem: it specifies a causal direction. The multiple regression model, an extension of the simple regression model, helps overcome the latter problem, as we will discuss next.

9.4 Items for Review

- Covariance
- Correlation coefficient
- Fisher's z transformation
- Confidence bounds on the correlation
- Hypothesis testing on the correlation
- Intercept and slope
- Error term
- Least squares criterion
- Estimators for intercept and slope
- Regression ANOVA table
- Standard errors of intercept and slope
- Regression hypothesis tests

9.5 Homework

Below is a small data set of 10 persons measured on 4 variables: life satisfaction, health, happiness, and education. Assume all variables are continuous.

Person	Satisfaction	Health	Happiness	Education
1	21	3	2	17
2	19	2	1	14
3	25	2	1	13
4	26	3	1	16
5	26	2	1	12
6	24	3	1	16
7	20	1	3	17
8	10	1	0	8
9	23	2	1	13
10	25	2	2	12

1. Some use life satisfaction and happiness measures as exchangeable, arguing that satisfaction and happiness are the same thing. Compute the correlation between

satisfaction and happiness and construct a confidence interval for it. How strong is the relationship? Use the confidence interval and hypothesis testing approach to answer this question.

2. Some argue that education influences health. Regress health on education and test the hypothesis that education and health are not related. Construct the ANOVA table and interpret all results.
3. Compute the correlation between health and happiness and construct a confidence interval for it.
4. Some argue that health limits one's ability to obtain more years of schooling. Regress education on health and test this hypothesis. Construct the ANOVA table and interpret. Given these results versus those in the previous question, can you say anything about the direction of causality?

Chapter 10
Introduction to Multiple Regression

In the previous chapters, we have focused on relationships between only two variables at a time. Most relationships that we are interested in social science research are more complicated, however, than can be understood with only bivariate analyses. In some cases, the relationship between two variables depends entirely on a third variable, and so a bivariate analysis can be misleading. In some cases, the relationship between two variables depends on, or operates through, additional variables. In this chapter, we will discuss multiple regression. In multiple regression analysis, a single outcome variable is modeled as a linear combination of as many additional variables as desired. Multiple regression is sometimes used simply to understand factors that are relevant to predicting an outcome, but it is also used in science to help establish that the relationship between two variables is a causal one and to understand complex relationships that simply cannot be understood with bivariate analyses. In this chapter, before introducing the details of multiple regression, we will discuss causal thinking in order to lay the foundation for seeing why multivariate analyses are necessary.

10.1 Understanding Causality

Although we have not discussed causality until now, the assumption that one of our goals in statistical analysis is often to establish causal relationships between variables has been implicit. We usually want to know whether group differences in means exist, because we may think the grouping variable *causes* the outcome variable. For example, we may be interested in examining whether there are gender differences in earnings because we believe that discrimination in the labor market is present, thus causing earnings differences. We may be interested in examining the correlation between education and earnings, because we may believe that education develops skills (or perhaps just credentials) that are valuable on the market and therefore rewarded with higher wages. We may be interested in examining differences in marital status distributions across regions of the country, because

S.M. Lynch, *Using Statistics in Social Research: A Concise Approach*,
DOI 10.1007/978-1-4614-8573-5_10, © Springer Science+Business Media New York 2013

we may believe that cultural practices vary across regions and make individuals more likely to marry, less likely to divorce, etc. Each of these examples implies a causal process in which the "independent variable," commonly denoted x, affects the "dependent variable," (or "outcome variable") commonly denoted y.

Causal arguments are ubiquitous in discussions among lay people. One only needs to look at comments on news blogs as people discuss virtually any topic. Is spanking a practice that enhances or discourages successful development in children? Does raising taxes increase or reduce growth? Does exercise reduce the risk of obesity and heart disease? Does having a strong safety net encourage dependency? While arguments on both sides of any debate often explicitly use causal language, rigorously establishing that a relationship between x and y is a causal one is incredibly difficult. First, defining causality itself is a difficult enterprise, one with which philosophers have struggled for literally millenia. Once defined, the process of determining whether the relationship between x and y meets the criteria for causality in social research is at least as difficult, because the process is riddled with the limitations of data and methods. For this reason, I have intentionally avoided discussing causality until this chapter. The previous chapters, as discussed above, have certainly alluded to the use of statistical testing as a means to peer into the window of causality, but I have intentionally avoided using the results of analyses to make causal claims, much as I have avoided claiming that the results of analyses prove any hypothesis to be true.

A very basic, initial problem with making causal claims can be seen by rediscussing the fallacy of affirming the consequent. Recall from Chap. 2 that the key reason that we cannot "prove" a hypothesis true is that alternate explanations may always exist that account for patterns we observe in data. Recalling the example in Chap. 2, I may claim that, if it rains today, my yard will be wet when I get home. When I get home, if my yard is, in fact, wet, I cannot conclude that my claim of rain is true, because my wife could have watered the lawn,[1] a water main could have broken, etc. This same rationale is true when we are seeking to make causal claims. For example, I could say: If A causes B, then I will find Z in my data (whatever Z is). Clearly, as per *modus tollens,* if I find that Z is not true in my data, I can conclude that A does not cause B. However, if I find Z, I cannot immediately conclude that A causes B. To do so requires that I meet a number of criteria, especially including most importantly that I rule out all alternate explanations.

What is Z? In other words, what should we observe if A causes B? The classic understanding of causality used in social science is that A is a cause of B if (1) there is some relationship between A and B, (2) the relationship is temporally ordered so that A precedes B in time, and (3) the relationship between A and B is not spurious.

The first requirement, correlation/association, is straightforward: If A causes B, then A and B must be associated in some fashion. If we think of A as an event, then when A happens, all else being equal, B must happen. If pushing someone

[1]In fact, my wife has informed me that she would never do so. We may have to talk about this, but doing so is beyond the scope of this chapter.

causes falling, then when I push someone, they should fall, all things considered. This requirement implies the second requirement: that the cause must precede the effect. Scientific views of causality do not allow for concepts like "destiny" in which effects produce causes: causes must come first.[2]

While establishing the first two requirements may seem straightforward, it is not necessarily easy to do so. First, two variables may be related to one another, but not linearly. That is, if the correlation coefficient is the measure you are using to capture the association between two variables, and the relationship is u-shaped, the correlation will be 0, even though a relationship clearly exists. Thus, it is important to properly model the shape of the relationship between two variables before appealing to *modus tollens* and *ruling out* a causal relationship, and multiple regression enables doing so, as we will discuss later in the chapter.

Second, it is not straightforward to establish temporal order. In cross-sectional data like the GSS in which all variables are measured simultaneously on individuals, we cannot establish which variable comes first except in rare cases. For example, parents' completed educational attainment almost certainly precedes respondent's educational attainment, and so the temporal order may be assumed. However, suppose we were interested in the relationship between health and happiness. Does having one's health make one feel happier, or does being happy make one feel healthier? Or, is the causal relationship possibly even reciprocal?

Having panel data—that is, data in which the same respondents are measured on more than one occasion—can help establish temporal order, but even with such data, temporal ordering is not always clear. For example, in examining the relationship between education measured at time 1 and health measured at time 2, we might assume that education precedes health. However, it may be the case that health prior to time 1 (time 0) is the cause of education at time 1. While it may be the case that education and health are both causes and effects of each other—so that health affects education, which then affects later health—more problematic is that health at time 0 may be the cause of both education at time 1 and health at time 2, and education is not a cause of health at all. This example brings us to the third requirement: non-spuriousness.

The word "spurious" in common language simply means false. In statistical usage, a spurious relationship between A and B is one in which a third variable, C, accounts entirely for their relationship. In other words, C is the cause of both A and B, and so the apparent relationship between A and B is a false one. A classic example of a spurious relationship is that between ice cream consumption rates (say gallons consumed per 100 people per week) and violent crime rates (say number of assaults reported per 100 people per week): the relationship is entirely explained by season (temperature). People eat more ice cream in warmer weather, and more violent crime is committed during warmer weather. Once temperature is taken into

[2]Teleological arguments—arguments in which actions in nature are purposeful, with some sort of predetermined endpoint that causes actions along the way to the endpoint—have long been rejected in science.

account, the relationship between ice cream consumption and crime vanishes. This example also suffers from a possible ecological fallacy: if the relationship were not spurious, is the implication that ice cream consumers are more likely to be criminals? As another, individual level example, consider the relationship between siblings' heights. Sibling heights are certainly correlated with one another, but the relationship is certainly not causal: the relationship is spurious because parents' height fully explains the relationship (see Davis 1985).

How do we rule out spuriousness as an explanation for the relationship between two variables of interest? There are two ways that researchers attempt to do so. One is the use of experimental methods, and the other is the use of statistical methods like multiple regression modeling.

10.1.1 The Counterfactual Model and Experimental Methods

To understand how experimental methods work, it is useful to define the counterfactual model of causality. At a most basic level, in order to demonstrate that some factor is a cause of something else, we must be able to show that the world, after receiving the cause, is different than it would have been without the cause. Consider a case in which we are trying to determine whether some new drug has an effect on reducing blood cholesterol. To demonstrate that the drug reduces cholesterol, we need to be able to show that an individual's cholesterol falls after taking the drug and that, counterfactually, his cholesterol would not have fallen had he not taken the drug, all else being equal—in other words taking vs. not taking the drug is the only thing that differs between the two conditions.

The counterfactual model is intuitively easy to grasp, and it should be employed at least as a thought experiment when thinking about causality. For example, consider the argument that gun control measures do not work, because we have some measures in effect already, and we still have a high level of gun violence. If we think counterfactually, we might ask: well, what would the level of gun violence be if all else were equal and we *didn't* have the measures in effect that we already have? Suppose we claim that toothpaste brand x prevents tooth loss, but we observe that, at age 70, a person who has used brand x for his entire adult life begins to lose his teeth. Does that imply that brand x does not prevent tooth loss? The counterfactual approach would ask: at what age would the person have begun losing his teeth if he *hadn't* used brand x?

While the counterfactual approach is an extremely useful way to think about causality, it is impossible in reality to implement. We cannot go back in time and see what our level of gun violence would have been had the gun control measures currently in effect had not been in effect. We cannot subtract 50 years of life off an individual to see the age at which tooth loss would have begun had he not used brand x. In short, it is impossible to observe a single individual, nation, or other unit of analysis in two states in which nothing differs except for some treatment of

Step 2 Group	Step 3 Pre-test	Step 4 Treatment	Step 5 Post-test	Change
R(T)	O_{it}	X	$O_{i(t+1)}$	$O_{i(t+1)} - O_{it} = \Delta_i$
R(C)	O_{jt}		$O_{j(t+1)}$	$O_{j(t+1)} - O_{jt} = \Delta_j$
Difference:	$\Delta_{(i-j)t} = 0$		$\Delta_{(i-j)(t+1)}$	$(\Delta_i - \Delta_j)$

Table 10.1. Illustration of a true experimental design applied to two individuals: i and j. $R(T)$ and $R(C)$ represent random assignment to treatment and control groups, respectively. Δ represents change or between individual differences.

interest. Suppose, for example, we measure an individual's cholesterol at time 1, administer the drug for a month, and then remeasure cholesterol at time 2. Then, we wait another month without the drug and remeasure cholesterol again at time 3. So, we have observed the person both with and without the drug. This strategy is a type of quasi-experimental design—a "time series design"—and is, in fact, commonly used in research (see Campbell and Stanley 1963). A key problem with this approach is that we have no way of knowing that nothing changed over the extended time period that may exacerbate or mitigate the estimated effectiveness of the drug. Perhaps the individual's diet changed. Perhaps the weather changed, and it somehow affects cholesterol. Perhaps the initial drug effect (assuming there was one) leads to a rebound increase in cholesterol so that the effect looks twice as large as it should. Perhaps the person is at a crucial age in which his/her cholesterol has begun a steady increase or decrease whether or not s/he took the drug. In short, the passage of time itself and factors associated with the passage of time makes the individual different under the two conditions (drug vs. no drug).

An alternative approach to obtaining a true counterfactual is to employ a "true experimental design." A true experiment involves at least two groups: a treatment group and a control group. In some cases, we include a placebo group instead of, or in addition to, the control group. A placebo group is a group that receives a fake treatment. First, we randomly select a sample from the population. Second, we randomly assign individuals to treatment and control (or placebo) groups. Third, we conduct pre-test measurement to establish baseline levels of the quantity of interest (like cholesterol level). Fourth, we provide the treatment to the treatment group and nothing to the control group (or a placebo to the placebo group). Fifth, after some passage of time, we remeasure the quantity of interest for both groups. The difference between the average change for the treatment group and the average change for the control/placebo group is the "causal" effect of the treatment.

Table 10.1 illustrates this process applied to two hypothetical individuals, i and j. Step 1—random sampling—is assumed. Step 2 is the process of random assignment (called randomization) to treatment and control groups. The second column shows the third step: the pre-test measurement for each individual at time t (O_{it} and O_{jt}). The third column shows the fourth step: the implementation of the treatment X to one, but not the other, individual. The fourth column shows the fifth step: the post-test measurement at time $t+1$ for each individual. The final column shows the difference within individuals from time t to $t+1$.

The bottom row of the table shows some important differences. The difference in the second column is the difference between the two individuals at baseline—the pre-test difference. The table shows that this difference is 0, and we will discuss why shortly. The difference in the bottom row of the fourth column is the post-treatment difference. Note that this is the difference *between* the two people and not a measure of change *within* the same person. Finally, the last column shows the within-individual change from pretest to posttest for both the treatment and control group, and the difference in these changes between the groups. This final difference is the causal effect of the treatment.

Under the counterfactual model, persons i and j would be the same person; however, as we discussed above, this is not possible: We cannot observe person i (aka j) both with and without the treatment. The best approximation to this ideal, instead, is to find two people who are *exactly* alike, give one the treatment and one the placebo (or no treatment), and observe the change for both as shown in the table. Obviously, it is impossible to find two identical people. However, if n is large enough and we *randomly assign* sample members to treatment and control groups, all factors that differentiate the members of the two groups will balance out, and pretest differences between treatment and control groups should be 0. This difference of 0 is for *all* characteristics, and not simply the quantity of interest. For example, suppose that our random sample consists of 50 men and 50 women. If we randomly select a person from this sample and assign him/her to the treatment group, then randomly select a second person and assign him/her to the control group, and continue this process until all sample members are assigned, we will end up with roughly 25 men and 25 women in each group. You can think of this process as being the same as the process of taking two independent random samples of some numeric characteristic from a single population: the two sample distributions should be similar to one another, in shape, mean, median, variance, etc. There will certainly be some slight difference, but if the samples are large enough, the differences will be minimal including on both characteristics you observe and characteristics you do not!

After the administration of the treatment, there are two sources of change from pre- to post-test in the treatment group: the passage of time and the implementation of the treatment. There is only one source of change for the control group: the passage of time. By subtracting the change for the control group from the change for the treatment group, we essentially eliminate the effect of time passage, and what is left is simply the effect of the treatment. Given that we have used a sample of individuals in both groups, the quantities in Table 10.1 should be means. Thus, we can place bars over the table quantities and change the subscripts to reflect groups rather than individuals. The result is that our experiment produces a quantity we can call the average treatment effect: $(\bar{\Delta}_T - \bar{\Delta}_C)$.

The process of randomization is fundamental to the validity of the true experimental design. Without randomization, we cannot completely rule out spuriousness, in part because we simply cannot measure all sources of it. A key problem is that, without randomization, we cannot know that the process of *selection* into a treatment group is ultimately responsible for differences we observe in some outcome between treatment and control groups and not the treatment itself. By selection, I mean that differences in pre-test characteristics may be responsible for the assignment to the treatment versus control group.[3] For example, consider the relationship between education (a treatment) and earnings (an outcome). These two variables are strongly related in data (e.g., in the 2010 GSS, the correlation is .38 for working age people). Furthermore, education almost certainly precedes earnings: people tend to finish schooling before entering the labor market. Can we conclude that education causes earnings, and therefore recommend either to individuals or to policymakers that people obtain more schooling? Unfortunately, although common sense may tell us we can, in reality we cannot do so on the basis of the evidence. We have not randomly assigned education to individuals, and so there are numerous alternate explanations that may render the strong correlation spurious—in this context, such spuriousness is called selection bias. Those with greater intelligence may be more likely to select themselves into higher levels of schooling *and* into more lucrative ventures. Those with greater motivation may be more likely to continue schooling and more likely to seek promotions and job changes to increase earnings. Those whose parents have greater wealth may be better able to afford higher education, and their parents may also have better connections to help their children obtain lucrative jobs upon graduation. In terms of Table 10.1, the pre-test differences that were 0 under randomization are not guaranteed to be 0 without it.

The problem of lack of randomization applies to the relationship between *any* variables we may be interested in examining. Suppose a nutrition supplement company is seeking to determine whether a new supplement they produce helps with weight loss. So, they advertise for volunteers to try their product. They weigh volunteers before and after using the product for 1 month, and they observe that the average weight loss among the volunteers was 10 pounds. Can they conclude that their product works? Of course not. Those who volunteered are almost certainly more motivated to lose weight than those who did not volunteer, and they may therefore be more likely to engage in additional behaviors to lose weight during the treatment period. Thus, selection into the study (and therefore the treatment group)—and not the product itself—is quite likely the cause of the weight loss. This study design, even if it were to have random selection, suffers from another serious problem: there is no control (comparison) group. In this design, given that there was only one treatment group, random selection into it is equivalent to randomization, with all non-selected population members being implicit controls. The key assumption is that weight change among the general populace was 0 over the study period. But what if, over the course of the study, a natural disaster restricted food availability to everyone, so that everyone in the population lost 10 pounds over the month? If no measurement of weight change among controls was made, the

[3]This can happen either because the respondent selects him/herself into a treatment vs. control group or because the experimenter does. For example, suppose an experimenter assigns sicker patients to the control group in an experiment evaluating how well a new drug works.

effect of passage of time (or factors, like famine, associated with it) cannot be ruled out as an alternate explanation for why those in the treatment group lost weight.

Despite the importance of randomization, it is an unobtainable ideal in most social science research. Many of the treatments (causes) that we are interested in investigating simply cannot be randomly assigned to individuals. We cannot, for instance, assign individuals to be poor so that we can evaluate their long-term health. We cannot assign teen pregnancy. We cannot force some people to exercise while excluding others from doing so. Instead, social scientists generally rely on survey (or similar) data in which treatments are observed but not randomly assigned, and only outcomes (i.e., post-test measures) are measured. For example, the GSS measures education and income simultaneously. Education may be the treatment of interest, and income is only observed *after* the treatment: we usually do not know what income was before schooling was completed.

In most social science research, therefore, we turn to the second approach to ruling out spurious explanations in order to isolate the effect of treatments of interest: statistical methods. There are, in fact, a number of statistical methods that have been developed—and continue to be developed—that can help us rule out spuriousness when dealing with non-experimental data. We will focus here only on multiple regression. Many other methods for evaluating causality rely on the basics of multiple regression modeling—indeed, multiple regression is often a part of more advanced methods—and so an understanding of it is fundamental.

Before turning to multiple regression, however, it is important to note that, while true experiments are usually seen as the gold standard for making causal claims, they are not without their limitations. First, single, causal claims do not exist in a vacuum: reality is more complicated. As mentioned above, health in early life may influence education, which may then influence later health. Thus, education is an *intervening* variable (or "mediator") in a multi-step causal process. Yet, it is difficult to evaluate more than one treatment in an experiment.

Second, aside from the difficulty with handling multiple, intertwined causal relationships, it is difficult to see how one could even begin to address how early life health influences later life health in an experiment. Experiments are usually of short duration for ethical and practical reasons and cannot generally be used to evaluate causal processes that unfold over the long term.

Third, experiments are usually small in scale and involve unnatural conditions, and it is unclear whether causal effects identified in them can always scale-up. For example, suppose it were possible to learn via an experiment that obtaining more education, in fact, does increase earnings. Would this finding still hold if everyone in a society increased his/her years of schooling? As another example, in research on the effects of one's neighborhood on various outcomes, some experimental work has shown that moving to a better neighborhood produces improvement in outcomes. But, is it feasible for everyone in society to move to a better neighborhood?

These limitations of experiments, as well as others, suggest that, at least in some cases, using statistical methods may be a better—or at least more realistic—strategy for answering research questions involving causal claims. Multiple regression is a key such method, and its strengths include that it can handle nonlinear relationships between variables, it can "control" out observed spurious threats to causality, and it can be used to attempt to disentangle direct and indirect causal relationships.

10.1.2 Statistical Control

In introducing multiple regression, let's consider one of the examples discussed in the previous section: the relationship between education and income. As mentioned above, the correlation between education and income is .38. Is this relationship causal? A respondent's current level of education almost certainly precedes his/her income, and the relationship between education and earnings is moderately strong. Yet, there may be alternate explanations for the relationship. For example, parental education may influence the respondent's educational attainment, and it may influence respondent's income for a variety of reasons, including that parents with greater education may be able to assist the respondent in choosing and obtaining a lucrative career path. Thus, the correlation between education and earnings may be spurious. In the context of Table 10.1, the implication is that pre-test differences are not 0, because respondents have not been randomized to education levels.

The correlations between parental education and respondent's education, as well as between parental education and respondent's income supports the view that respondent's educational attainment is not randomly assigned. The correlation between parent's education and respondent's education is .48, which is even larger than the correlation between respondent's education and income. Figure 10.1 illustrates the concern. In the figure, there is an arrow from parents' education to both respondent's education and respondent's income. The question is: once these two relationships are "controlled," is there any relationship remaining between respondent's education and respondent's income?

Multiple regression allows us to statistically control for parent's education to examine the remaining relationship between respondent's education and income. Before showing the process of control in multiple regression, however, we should discuss the meaning of "statistical control." Under randomization, all pre-test differences between the control and treatment groups are 0: random assignment ensures that there are no differences in any characteristic. In observational data, there is no guarantee, but statistical control is used to adjust pre-test differences so that the treatment and control groups are equal. A simple, albeit inefficient method of statistical control in the current example would be to examine the relationship between respondent's education and income among respondents with the same level of parental education. In other words, although we cannot assign individuals to the treatment and control groups (i.e., higher versus lower educational attainment) to balance the groups on pre-test characteristics, like parental education, we can hold

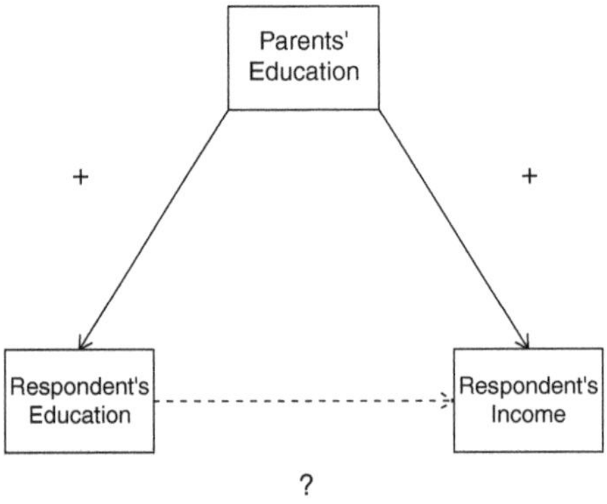

Fig. 10.1. Path model for the relationship between parents' education, respondent's education, and respondent's income.

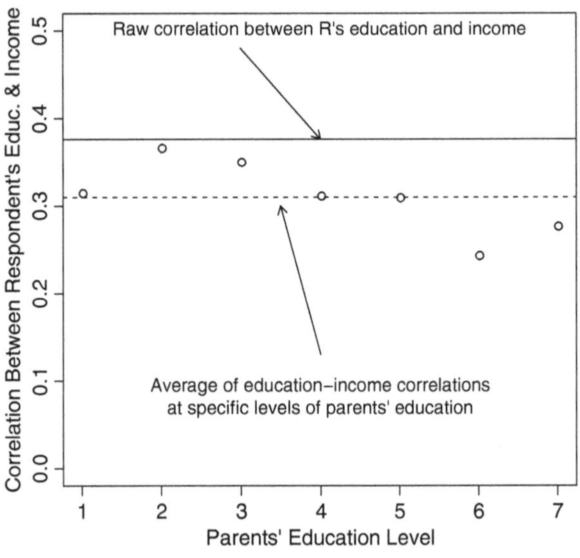

Fig. 10.2. Correlations between respondent's education and income for the sample (solid line), by level of parent's education (dots). Dashed line is the average of correlations at different levels of parent's education (levels of schooling discussed in text).

these characteristics constant by limiting our analyses to those with similar values on such characteristics. Put another way, we can manually balance respondents on parental education.

Figure 10.2 shows the results of following such a strategy. The solid horizontal line represents the correlation between respondent's education and income in the entire sample (.38). The dots are the correlations between respondent's education and income for persons with different levels of parental education. Parental education was coded into seven categories as follows: 1 for 0–7 years of schooling, 2 for

8 years of schooling, 3 for 9–11 years of schooling, 4 for 12 years of schooling, 5 for 13–15 years of schooling, 6 for 16 years of schooling, and 7 for more than 16 years of schooling. As the figure shows, while the raw (bivariate) correlation between education and income is .38, the correlation between education and income among persons with the same level of parental education tends to be much lower. The average correlation between education and income across all levels of parental education is .31. This correlation is 18 % smaller than the original correlation and illustrates that, once parents' education is controlled—that is, for respondents at comparable levels of parental education—the relationship between education and income is somewhat smaller than the raw correlation.

Why is the raw correlation between education and earnings larger than the correlations between education and earnings at each level of parental education? The answer is that respondents with lower education tend to have parents with lower education (and vice versa), and parental education influences the assignment of education level to the respondent *and* has a residual relationship with respondent's income. Ignoring parental education therefore consolidates all of the effect of parental education on respondent's income into respondent's education.

10.2 Multiple Regression Model, Estimation, and Hypothesis Testing

Multiple regression is more efficient at control than using the disaggregation method I just demonstrated. The multiple regression model extends the simple regression model to handle additional independent variables. The extended model is:

$$y_i = \beta_0 + \beta_1 x_{i1} + \beta_2 x_{i2} + \beta_3 x_{i3} + \ldots + \beta_k x_{ik} + e_i. \tag{10.1}$$

Notice that in this equation (compared to the simple linear regression model), α has been replaced by β_0, β is now more than one slope and is now subscripted, and the x variables are double-subscripted to reflect the measurement of multiple variables per individual. Notational differences aside, the multiple regression model is a straightforward extension of the simple regression model. Essentially, whereas in simple linear regression we estimated the *line* corresponding to the relationship between an x and a y, multiple regression extends the geometry to multidimensional *planes,* capturing the "tilt" of these planes in the different dimensions.

Estimation of this expanded model follows the same criteria as the simple regression model—finding the values of the parameters that minimize the sum of the squared errors—but the solution involves linear algebra and is:

$$\hat{\beta} = (X^T X)^{-1}(X^T Y). \tag{10.2}$$

Observe that I have replaced β with $\hat{\beta}$ (or b); I have done so because this is the *estimated* "vector" of slopes (or "regression coefficients"). The $(X^T X)^{-1}$ matrix (read: "x transpose x inverse") is akin to the denominator of the solution for β in the simple regression model $(\sum(x - \bar{x})^2)$; the $(X^T Y)$ vector (read: "x transpose y") is akin to the numerator of the solution for β in the simple regression model $(\sum(x - \bar{x})(y - \bar{y}))$. Just as in the simple regression model in which α and β are the best estimates of the population regression parameters, the estimates of β in the multiple regression are also the best estimates of the simultaneous relationships between each x variable, controlling on (or "net of") all other x variables in the model.

Given the size of most data sets in social science research, as well as the relative complexity and tediousness involved in computing the estimates of the regression coefficients, we will not dwell on the estimation of the model parameters and standard errors. In general, researchers use statistical software packages to perform the estimation. So, we will focus on hypothesis testing, understanding the model, and expanding the model's capabilities.

The ANOVA table for the multiple regression model looks exactly like the one from the simple regression model, with one exception: the degrees of freedom for the regression sums of squares is equal to the number of independent variables in the model (technically, it is the number of parameters, including the intercept, minus 1). The rest of the ANOVA table looks just as before, and so we will not repeat it here. In most tabular presentations of multiple regression model results, the only elements of the ANOVA table that are displayed include the model R-squared and possibly the model ANOVA F statistic.

The F test from the ANOVA table is now an omnibus test that tests whether *any* of the independent variables has a linear relationship with the outcome variable, and the model's R^2 reflects the proportion of the total variance in y that is explained by the linear combination of all the independent variables in the model. Ideally, if our model fits the data well, R^2 will be high, and the F test will be statistically significant. Technically, however, R^2 is a measure of both model fit and the extent of noise in the outcome: A model may fit very well, but y may simply have considerable noise built into it. So, a low R^2 should not necessarily be considered evidence that the model fits poorly.

Although the F test is important, we are usually interested in whether specific variables are related to the outcome. Thus, the main statistical tests that are of interest in the multiple regression model are t-tests on the parameters, just as with simple regression. As we discussed in the previous chapter, the logic of the test is as follows. If a particular x affects y, then we would expect the corresponding β to be non-zero. Put another way, if x and y are unrelated, then we would expect the best estimate for y to be $\bar{y}$, which does not depend on x. Thus, the slope (β) should be 0. The t-test to determine whether β is 0 can therefore be conducted for each parameter just as we did in the previous chapter for simple linear regression:

$$t = \frac{\hat{\beta} - 0}{SE(\hat{\beta})}. \tag{10.3}$$

Variable	Model 1	Model 2	Model 3
Intercept	−13.8(.75)***	8.7(.54)***	−16.1(.76)***
Education	3.1(.05)***		2.7(.06)***
Parental Educ.		1.7(.04)***	.67(.05)***
R^2	.14	.07	.15

$*p < .05, **\ p < .01, ***p < .001$

Table 10.2. Regression of income on respondent and parental education (GSS data 1972–2010, ages 30–64, $n = 20,409$; coefficients and (s.e.) presented)

This t-test is interpreted the same way as the t-tests we have used before: it provides us a measure of the probability we would observe the sample (regression coefficient) we did if, in reality, there is no relationship between x and y in the population. If the test value is large, the probability of observing the coefficient we did would be small under the assumption that $\beta = 0$ in the population. Thus, a significant t-test is usually considered evidence that x is related to y.

In order to illustrate the multiple regression model, I return to the example discussed previously involving parental education, respondent education, and respondent income. For this example, I use data from the 1972–2010 GSS. Only persons ages 35–64 with non-zero income were included. Education for both parents and respondents is measured in years of schooling (from 0 to 20). Parental education is measured as the maximum years of schooling for those with data on two parents. Finally, income is measured in \$1,000s in real (2010) dollars.

Table 10.2 shows the results of three regression models using the data. In the first model, income was regressed on respondent's education (y is always regressed on x), and the parameter estimate (b_1) was 3.1, meaning that, on average, each year of schooling is associated with an increase of \$3,100 in income. In the second model, income was regressed on parents' education, and the parameter estimate (b_1) was 1.7. That is, each year of parental education is associated, on average, with a \$1,700 increase in income. Finally, Model 3 includes both independent variables. In this model, the parameter for respondent's education (b_1) is 2.7, and the parameter for parental education (b_2) is .67. Note that I have used the word "increase" although the model really simply tells us that those with more education have more income on average. The use of the word "increase" implies causality, and it implies within-individual change, neither of which we have demonstrated. This grammar usage is common, albeit not exacting.

The asterisks in the table indicate that the t-tests on all coefficients were statistically significant at the .001 level, meaning that there is a tiny probability that the estimated coefficients would be the magnitude they are if they were each 0 in the population. Substantively, education and parental education both appear to be related to respondent's income.

The results of Model 3 can be used to estimate (or predict) respondents' incomes based on their education level and the education level of their parents, following Eq. 10.1. For example, suppose we are interested in estimating income for a person with 12 years of schooling whose parental education is 16 years:

$$\hat{y}_i = b_0 + b_1 E_i + b_2 PE_i \tag{10.4}$$

$$= -16.1 + 2.7E_i + .67PE_i \tag{10.5}$$

$$= -16.1 + 2.7(12) + .67(16) \tag{10.6}$$

$$= 27.02 \tag{10.7}$$

Thus, persons with those values of education and parental education make, on average, about $27,000 in income.

In Model 3, both parameters are smaller than their counterparts in the first two models. Why? This change in magnitude illustrates the concept of control in the multiple regression model: approximately $1 - 2.7/3.1 = .13$ (13 %) of the relationship between respondent's education and income is due to parental education. Put another way, at comparable levels of parental education, there is a $2,700 difference in respondent's income on average. Compare the proportion of the relationship explained here with the reduction in the correlation between respondent's education and income shown in Fig. 10.2. The proportions are close, albeit not identical, but their similarity illustrate the notion of control: When a variable is controlled in multiple regression, it means that we are holding that variable constant to evaluate the relationship between other variables and the outcome of interest.

10.2.1 Total, Direct, and Indirect Association

We can see this idea of control in more detail if we define total, direct, and indirect association. Reconsider the path diagram shown in Fig. 10.1. In that diagram, we assumed that parents' education precedes both respondent's education and respondent's income, and the question is the extent to which respondent's education and income are related after controlling on parents' education. The regression coefficients shown in Table 10.2, with some adjustment, can help us evaluate the total association between education and income, and the indirect association of parental education with income *through* respondent's education.

The adjustment we need to make to the raw regression coefficients shown in the table is to *standardize* them. We have seen standardized variables before in previous chapters: the z score is a standardized score. It tells us the number of standard deviations a value of some variable is from its mean. In the regression context, a raw (i.e., unstandardized) coefficient tells us the expected change in y for a one-unit change in x. This change is measured in the raw metric of the variables involved, as discussed above. A standardized regression coefficient, in contrast, tells us the expected standard unit change in y per standard unit change in x. In other words, how much does a one z score unit change in x influence a z score change in y? The computation of a standardized regression coefficient is straightforward:

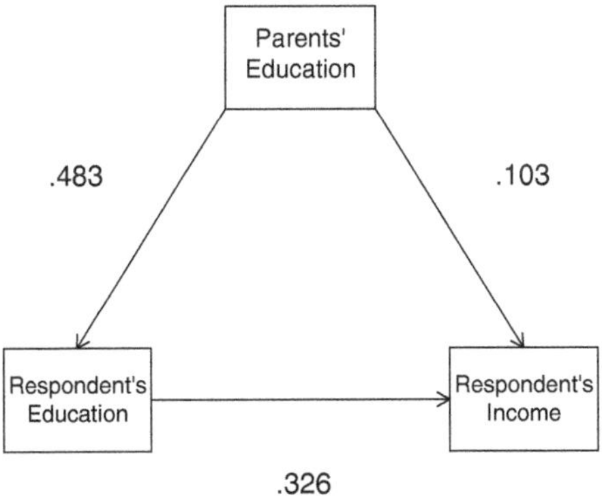

Fig. 10.3. Path model for the relationship between parents' education, respondent's education, and respondent's income with standardized coefficient estimates (GSS data from regression example)

$$\hat{\beta}_s = \left(\frac{s_x}{s_y}\right)\hat{\beta}. \tag{10.8}$$

In other words, we can convert to a standardized metric by simply multiplying the raw coefficient by the ratio of the standard deviation of x to the standard deviation of y.

The total association between x and y can be decomposed into the direct association between x and y and the indirect associations between x and y *through* other variables, using the following, simple identity:

$$Total = Direct + Indirect. \tag{10.9}$$

The indirect associations are simply the products of coefficients that comprise the pathways from x to y through other variables. For example, Fig. 10.3 revises Fig. 10.1 to include standardized regression coefficients.[4]

As the figure shows, the standardized association between parental education and income is .103, and the standardized association between respondent's education and income is .326. The standardized association between parental education and respondent's education was obtained via a regression model not shown here and was .483. Finally, two additional standardized estimates were obtained. One estimate was for Model 1 in Table 10.2: the total association between respondent's education and income was .376. The other estimate was for the total association between parental education and income: .261. That is, total associations between one variable and another are obtained from a simple linear regression model.

[4]Total, direct, and indirect effects do not need to be in a standardized metric. However, I use standardized coefficients here so that the relative magnitudes of the effects are more apparent.

Let's consider the relationship between parental education and respondent's income. The total association between parental education and income is .261, and as shown above, it can be decomposed as:

$$T = D + I \tag{10.10}$$

$$.261 = .103 + (.483)(.326). \tag{10.11}$$

Thus, of the total association between parental education and respondent's income, 61 % (1 − .103/.261) of it is accounted for by respondent's education. Thus, respondent's education is called an *intervening variable* and is said to *mediate* much of the relationship between parental education and respondent's income. Put another way, respondent's education largely explains why parental education is strongly related to respondents' income: parents with high levels of schooling tend to produce children who obtain high levels of schooling and parlay it into higher incomes.

Now let's consider the total association between respondent's education and income:

$$T = D + I \tag{10.12}$$

$$.376 = .326 + (.483)(.103) \tag{10.13}$$

Here, about 13 % (1−.326/.376) of the relationship between respondent's education and income is due to parental education. In other words, part of the relationship is spurious, but only a small part.

Notice that, although the path diagram has a directed arrow from parental education to respondent's education, the standardized coefficients are agnostic with respect to direction of causality. Thus, the decomposition of total association into direct and indirect is useful for either evaluating the extent to which a relationship between two variables is spurious *or* understanding mediating processes that explain why one variable has a relationship with another.

Throughout this section thus far, I have used the term "association" rather than "effect." The reason is that the methodology of decomposing total associations into direct and indirect associations is not in and of itself a method for establishing causal relationships. Nonetheless, when we construct path models with arrows delineating the direction we believe relationships "flow" between variables, we are implying causal pathways. Thus, we often use causal terminology in our discussion of such models. I will do so subsequently largely out of convenience, but keep in mind that the relationships I am discussing may not necessarily be truly causal ones.

10.2.2 Suppressor Relationships

While it is clearly important to rule out spuriousness, and thus include potential "upstream" variables in a regression model—that is, variables we think are causally prior to our variable of interest—is it important to include variables in the model

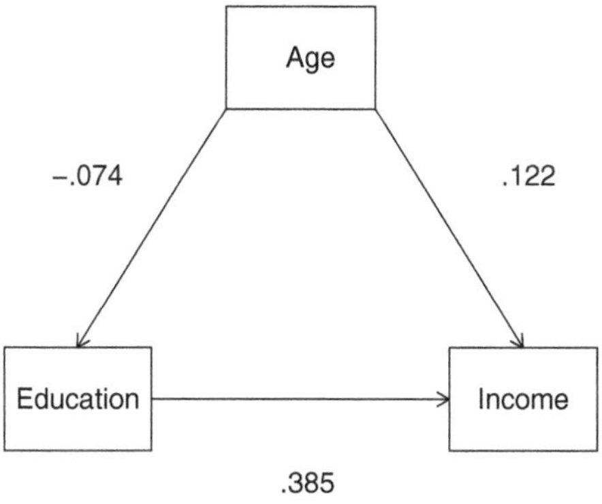

Fig. 10.4. Path model for the relationship between age, education, and income

that are "downstream," that is, intervening variables between the two variables of interest? For example, if we are really interested in the effect of parental education on income, is it necessary to control on respondent's education? In some sense, it may seem like it is unnecessary to do so, since respondent's education is an effect along the way from the ultimate cause of interest (we sometimes call such intervening variables "proximate causes," with ultimate causes called "distal causes"). However, the total effect of x on y can be misleading. In the current example, all the relationships between variables are positive: parental education is positively related to respondent's education and income, and respondent's education is positively related to income as well. Yet, this type of structure is not always present. Sometimes, a distal cause has a positive effect through one intervening variable, but a negative effect through another. Thus, the total effect may appear to be small or even zero.

We call this type of relationship a suppressor relationship, and ignoring it can produce misleading inferences. In order to illustrate suppression, consider the relationships between age, education and income as shown in Fig. 10.4. The figure contains the standardized coefficients needed to decompose the total effect of age on income into direct and indirect effects through education, obtained from the same GSS sample used in the previous example.

We might expect age to have both positive and negative indirect effects on income, in part because age represents a combination of factors. First, at the individual level, age represents the passage of time—maturation. With aging comes a number of advantages that should increase income, like increased work experience and increased savings and investments yielding higher returns, including simply interest.

At the same time, however, recall that, in cross-sectional data, we do not actually observe individuals aging: we observe differences between people at different ages at a single point in time. While such differences might include changes that occur with aging, they also include changes that differentiate *birth cohorts* from one another. One large change that has occurred across birth cohorts in the U.S. and other Western countries is increases in educational attainment. For example, in the

Variable	Model 1	Model 2	Model 3
Intercept	18.1(.81)***	−13.8(.75)***	−28.7(1.1)***
Age	.24(.02)***		.32(.02)***
Education		3.1(.05)***	3.2(.05)***
R^2	.01	.14	.16

$*p < .05, **p < .01, ***p < .001$

Table 10.3. Regression of income on respondent age and education (GSS data 1972–2010, ages 30–64, $n = 20,409$; coefficients and (s.e.) presented)

GSS data, the mean years of schooling for those between 30 and 40 in the 1970s was 13.1 years. For those between 30 and 40 in the 2000s, the mean was 14.1, a full year's difference. As we've seen, those with greater educational attainment tend to have higher incomes. Taken together, then, age is negatively related to education, producing a negative indirect effect of age on income through education:

$$T = D + I \tag{10.14}$$

$$= .122 + (-.074)(.385) \tag{10.15}$$

$$= .094. \tag{10.16}$$

Notice that the total effect of age is actually smaller than the direct effect: this result occurs because the total effect of age on income contains both negative and positive effects. Table 10.3 shows the same result in the raw metric of the variables. In Model 1, the effect of age is .24, meaning that each year of age is associated with an increase of $240 in income. Model 2 is the same as in the previous example. In Model 3, with both age and education included, the coefficient for age is .32, some 33 % (.32/.24) larger than in Model 1. In short, even though education is a mediator of the relationship between age and income, it was important to include education in the model, because the total effect was suppressed by age's negative relationship with education.

Also realize again, that, if we were primarily interested in the relationship between education and income, it would have been important to control on age because education's effect was also suppressed by age: once age is controlled, the relationship between education and income appears larger. The reason is that age increases income, perhaps due to work experience, but older cohorts have less education, making education's total effect seem smaller.

Once again, despite the fact that we have spent much of the last few pages discussing "effects" of age, parental education, and respondent's education on income, can we really claim that we have established that these relationships are causal? The use of the word "effect" implies that we are assessing causal relationships, but such is the limitation of the English language: it is much more efficient to use this terminology than a more accurate one. Yet there are several reasons why we cannot conclude that the "effects" we have found are the true causal effects.

First, in each example, we have only controlled on one variable prior to the one of interest. If our interest is in the causal effect of education on incomes, in each example we have only controlled on one prior variable. In the first example, we only controlled on parental education; in the second, we only controlled on age. Yet we have shown in these two examples that both age and parental education predict respondent's education, so both should be controlled simultaneously. Indeed, in reality, there are a number of variables that should be controlled in order to meet the goal shown in Table 10.1 of making pre-test differences between the "treatment" and "control" groups (those with more vs. less education) zero. Sex and race both certainly precede respondent's education, and both certainly influence income through mechanisms other than educational attainment. For example, there are notable differences in the *type* of education men and women choose (i.e., career paths) that influence income beyond the simple number of years of schooling each gender obtains. Men are more likely to obtain 4-year college degrees in engineering than women, and women are more likely to obtain 4-years degrees in education. These professions, while requiring the same number of years of schooling, pay substantially differently. Thus, sex and race should be controlled. Countless other variables should be controlled as well, like region, religious affiliation, and others. In other words, our simple, two-variable multiple regression models are insufficient. A casual examination of social science articles, in fact, shows that most regression models include at least half a dozen controls. Unfortunately, we can *never* control on every possible spurious threat, because many variables are unobserved: surveys rarely measure enough variables to do so. Furthermore, many spurious threats are *unobservable,* like motivation. Thus, regression modeling by itself cannot accomplish what randomization can, and we should always be hesitant to make causal claims from such models.

Second, the multiple regression models I have shown so far assume the relationships between all variables in the model are linear. In many cases, this assumption may be reasonable, but in probably many more cases, it isn't. For example, the relationship between age and income is certainly not linear across all of adulthood. Although I have restricted the data to persons of working ages (30–64) in the preceding examples, it is not clear that the relationship between age and income is linear even in that age range. At some age, there is a tradeoff between what experience adds to a worker's value to an employer vs. what knowledge of new technology adds. Thus, we might expect that income increases across age to a point, but then potentially stagnates or even decreases after. Determining the true causal effect of age, then, requires that we be able to model nonlinear relationships between the purported cause and effect.

Third, in addition to the assumption that all relationships are linear in the multiple regression models presented thus far, we have also assumed that all relationships are *additive*. That is, for example, we have assumed that the relationship between age and income is the same regardless of the respondent's level of education. It is possible that the relationship between age and income is not the same across all levels of education. The relationship between age and income may follow one pattern for those with PhDs, say, while following another pattern altogether for those without high school diplomas. This possibility necessitates that we be able to model *interactions* between independent variables.

In the next three sections, I discuss how we can incorporate variables like sex and race, which are fundamentally non-numeric variables, into the multiple regression model; how we can incorporate nonlinear relationships into the model; and how we can incorporate interactive relationships into the model.

10.3 Expanding the Model's Capabilities

The multiple regression model would be of relatively little use in social science if it were limited to modeling linear relationships between continuous, numeric variables only. Many, if not most of the variables we use in social science research are not numeric, and the relationships between variables that our theories specify are often nonlinear. Fortunately, the multiple regression model is quite flexible and can accomodate nonlinear relationships and variables measured other than continuously. The one limitation that remains is that the outcome variable y must be continuous (and $y|x$ must be normally distributed; put another way $e \sim N(0, \sigma_e^2)$). When this assumption is violated, alternate models are needed (but discussing them is beyond the scope of this book).

10.3.1 Including Non-continuous Variables

The most useful extension of the regression model is the ability to incorporate noncontinuous covariates (predictors, x). If we are interested in race, sex, or other group differences in some y, we can incorporate indicator variables—also called "dummy variables"—into the model. For example, suppose we were interested in examining race differences in income in the GSS data. Race is measured at the nominal level (1 = white; 2 = black; 3 = other), and therefore the numeric codes assigned to racial categories in the data are meaningless. However, for regression analyses, we can construct a pair of variables indicating whether a respondent is black (=1) (or not = 0) and whether a respondent is an "other" race (=1) (or not = 0), and we can include those dummy variables in the regression model as predictors:

$$Income_i = \beta_0 + \beta_1 \times Black_i + \beta_2 \times Other_i + e_i. \quad (10.17)$$

If we estimate this model, we obtain:

$$E(Income) = 29.21 - 5.52 \times Black + 2.91 \times Other. \quad (10.18)$$

The interpretation of these coefficients is straightforward. If we were interested in the predicted value of family income for blacks, we simply insert Black = 1 and Other = 0 and compute the expected value:

Variable	Model 1	Model 2
Intercept	29.21(.19)***	−13.25(.76)**
Education		3.08(.05)***
Black	−5.52(.52)***	−3.12(.48)***
Other	2.91(.80)**	3.96(.74)**
R^2	.01	.14

* $p < .05$, ** $p < .01$, *** $p < .001$

Table 10.4. Regression of income on race and education (GSS data 1972–2010, ages 30–64, n = 20,409; coefficients and (s.e.) presented)

$$E(Income) = 29.21 - 5.52 \times (1)2.91 \times (0) = 23.69. \qquad (10.19)$$

If we are interested in the predicted value of income for persons of other races, we insert Black = 0 and Other = 1 and compute the expected value:

$$E(Income) = 29.21 - 5.52 \times (0)2.91 \times (1) = 32.12 \qquad (10.20)$$

Finally, if we are interested in the predicted value of income for whites, we insert Black = 0 and Other = 0. Doing so will leave us with only the intercept—the expected value of income for whites is 29.21. This latter finding shows that we only need to include $k - 1$ dummy variables for a variable that initially had k categories: The model intercept represents the value when all the other dummy variables are 0. The omitted group is therefore called the "reference" group, because the coefficients for the dummy variables represent how much the mean for the group represented by the dummy variable differs from the mean for the reference group. For example, in the example above, blacks have, on average, $5,520 less income *than whites*, while others have $2,910 more income *than whites*.

A couple of notes are in order regarding the use of only dummy variables in a regression model. First, the t statistic that you obtain for the regression coefficient will be identical to the t statistic you obtain when conducting a simple independent samples t test for the two groups (assuming you only have one dummy variable in the model). Second, one of the standard parts of the output of regression model software is an ANOVA table of the regression results. This ANOVA table will be identical to that which would be obtained if you performed the equivalent ANOVA. Indeed, with only dummy variables in the model, the coefficients for the dummy variables simply reproduce the mean of y for each group.

We can include both dummy variables and continuous variables in our models. Extending the example above, if we include education as a predictor of income along with race, we obtain the results shown in Table 10.4. We used the coefficients from Model 1 above in showing the interpretation of the dummy variable coefficients. Model 2 shows that education increases income (each additional year of schooling increases expected income by $3,080), and that, once education is controlled, the coefficients for the race dummy variables change. The dummy variable coefficient for blacks decreases in magnitude by $1 - 3.12/5.52 = 43\,\%$, while the dummy variable coefficient for those of other races *increases* in magnitude by $3.96/2.91 = 36\,\%$. Substantively, we can use the concepts of direct, indirect, and total association to discuss the implications of these changing coefficients. For blacks, the total association from Model 1 was negative: blacks make less income than whites. In Model 2, once education is controlled, this negative association decreases (shrinks) toward 0. Given that education has a positive relationship with income, the relationship between the black dummy variable and education must be negative—i.e., blacks have less education than whites on average—in order for the direct association to be reduced from a larger to a smaller negative number. Thus, if blacks had comparable education to whites, their incomes would be higher, and the income gap would be smaller. Put another way, 43 % of the black-white difference in income is due to education differences between these racial groups, with blacks having less education than whites on average.

The change in the coefficient for persons of other races is slightly more difficult to consider, because the coefficient increases away from zero once education is controlled. Again, we know that the relationship between education and income is positive; thus, the relationship between the other race dummy variable and education must be negative, in order for the direct association to increase. Here, education differences between whites and those of other races, with whites having more education on average, are suppressing the "other" advantage in income. In other words, if persons of other races had average educational attainment comparable to whites, the income difference between whites and others would be 36 % larger than it already is.

10.3.2 Statistical Interactions

In the discussion regarding the inclusion of dummy variables, we implicitly assumed that the relationship between education and income was the same for each race. Specifically, consider the prediction equation from Model 2 of Table 10.4:

$$E(income) = -13.25 + 3.08 \times Education - 3.12 \times Black + 3.96 \times Other \tag{10.21}$$

For each race, the prediction equation reduces to:

$$E(income)_W = -13.25 + 3.08 \times Education \tag{10.22}$$

$$E(income)_B = (-13.25 - 3.12) + 3.08 \times Education \tag{10.23}$$

$$E(income)_O = (-13.25 + 3.96) + 3.08 \times Education. \tag{10.24}$$

As these equations show, there are race differences in the expected value of income at each level of education—the intercept—but the return for each year of education (slope) is the same—$3,080—for each race. The result is three parallel prediction lines.

Variable	Model 1	Model 2
Intercept	−13.33(.79)**	−13.57(.84)***
Education	3.09(.06)***	3.10(.06)***
Black	−3.11(.48)***	−1.22(2.27)
Black*Education		−.14(.17)
R^2	.14	.14

* $p < .05$, ** $p < .01$, *** $p < .001$

Table 10.5. Regression of income on race and education (GSS data 1972–2010, ages 30–64, $n =$ 19,440; coefficients and (s.e.) presented)

Often, the assumption of parallel prediction lines across subgroups in a population is an unreasonable one, or it may be an assumption our theory/hypothesis challenges. For example, with regard to race, education, and income, one may argue that blacks are discriminated against in the workforce so that each year of education does not produce comparable income returns compared to whites. In order to capture differences in the effect of one variable across subgroups of the population, we can construct "statistical interactions" and include them in our model. In this example, constructing an interaction to capture differential returns to education for blacks involves creating a new variable that is the product of the black dummy variable and education. To simplify matters, let's examine only blacks and whites and create an interaction between black and education. The prediction equation becomes:

$$E(income) = b_0 + b_1 education + b_2 black + b_3(black * education). \quad (10.25)$$

For whites, then, the prediction equation is:

$$E(income) = b_0 + b_1 education, \quad (10.26)$$

while, for blacks, the prediction equation is (after setting the "black" dummy variable to 1):

$$E(income) = (b_0 + b_2) + (b_1 + b_3) education. \quad (10.27)$$

This result shows that whites and blacks differ in both intercept and slope for the effect of education. Table 10.5 shows the results of two models: one without a statistical interaction and one with the interaction.

Given the results shown in the table, we have the following empirical prediction equations:

$$E(income)_W = -13.57 + 3.10 \times education \quad (10.28)$$

$$E(income)_B = (-13.57 - 1.22) + (3.10 - .14) \times education \quad (10.29)$$

$$= -14.79 + 2.96 \times education \quad (10.30)$$

$$(10.31)$$

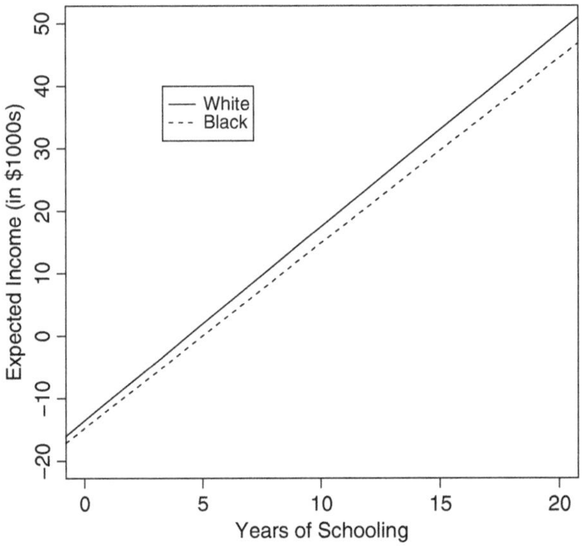

Fig. 10.5. Expected income by years of schooling and race (GSS data 1972–2010, ages 30–64)

From these results, it appears not only that blacks have lower incomes than whites at all levels of education, but also that each additional year of schooling nets less additional income for blacks relative to whites as well. Figure 10.5 displays these prediction equations graphically. As the figure shows, the prediction lines for blacks and whites are relatively close together at the lowest level of education (0 years), and the gap between the two racial groups expands across years of schooling. However, as the table indicates, in Model 2, neither the black dummy variable coefficient nor the interaction term is statistically significant. Thus, the main effects only model (i.e., the model without the interaction effect) is the better model for these data.

Interactions are not limited to two variables, nor are they restricted to a dummy and a continuous variable. However, interactions between continuous variables and three-way or higher order interactions are complex in interpretation and are beyond the scope of this book.

10.3.3 Modeling Nonlinear Relationships

Sometimes we are interested in modeling nonlinear relationships between variables. For example, in our last model predicting income as a function of race and education, we found that predicted income for persons with less than about 5 years of schooling was *negative*. Negative incomes do not occur with any regularity in reality, and so a model that predicts negative values of income may be unrealistic. Figure 10.6 shows the actual pattern for mean income by years of schooling in the

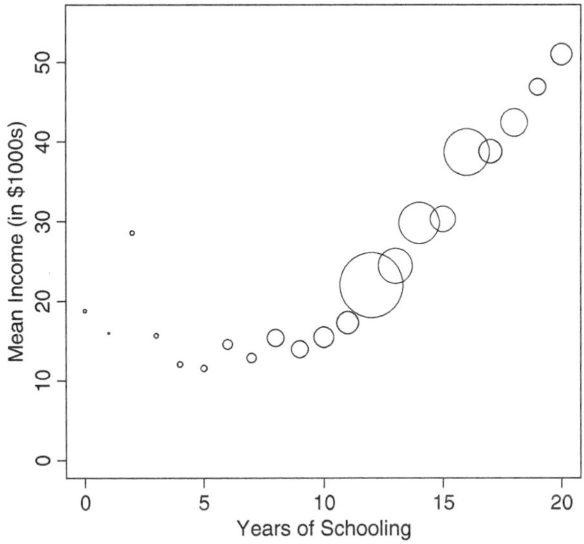

Fig. 10.6. Mean income by years of schooling with values weighted by sample size (GSS data 1972–2010, ages 30–64)

GSS data we have been using. As the figure shows, the relationship between years of schooling and income appears linear from about 8 years of schooling upward, but the relationship seems to bottom-out and perhaps even curve upward as education decreases from 8 to 0 years. Note that the plotting characters in the figure have been adjusted to reflect the number of observations at each level of schooling. As the figure shows, there are relatively few person in the sample at the lowest levels of schooling, and this explains why the model estimates a line that falls below 0 around 5 years of schooling: the estimates are driven by the much larger proportion of the sample at higher levels of schooling.

How can we capture this inverted curvilinear, or even u-shape pattern in a linear regression model? There are a variety of ways to capture nonlinearity, including applying nonlinear transformations to y variables like the logarithm. Indeed, the logarithmic transformation is a common one for income for several reasons, including that it reduces the right skew of the distribution. We will not illustrate the log or other transformations here; instead, we will focus on one particular class of models for capturing nonlinearity: using polynomial terms for x. This approach is called "polynomial regression."

Recall from algebra (and Chap. 5) that a general equation for a parabola is:

$$y = a(x - h)^2 + k, \qquad (10.32)$$

where (h, k) is the vertex, and a determines the breadth/curvature and direction of the opening of the parabola. If a is positive, the parabola is u-shaped; if a is negative, the parabola is inverted. If a is small, the parabola is wide; if a is large, the parabola is narrow.

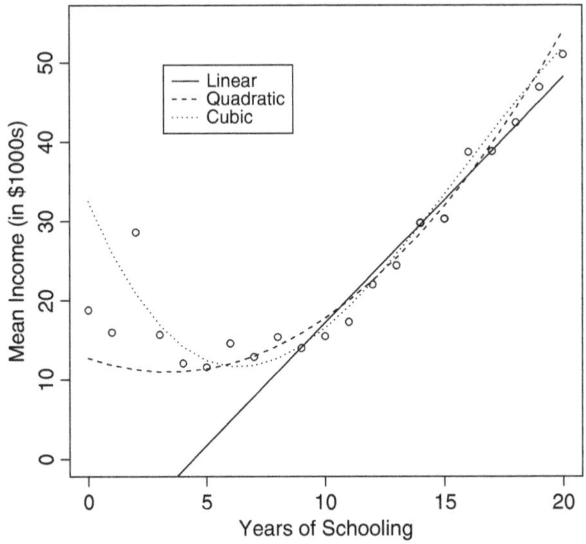

Fig. 10.7. Mean income by education with linear, quadratic, and cubic fits superimposed (GSS data 1972–2010)

If we expand the quadratic term in this formula, we obtain:

$$y = ax^2 - 2axh + ah^2 + k, \tag{10.33}$$

and if we then let $b_0 = ah^2 + k$, let $b_1 = -2ah$, and let $b_2 = a$, we obtain:

$$y = b_0 + b_1 x + b_2 x^2. \tag{10.34}$$

This result suggests that, if we include a second variable in our regression model—namely, x^2—we can capture a parabolic shape with the linear regression model. x^2 is constructed simply by taking the original x variable, squaring it, and saving this quantity as a new variable that we then include in the multiple regression model.

This result can be extended to higher order polynomials in order to capture more complex nonlinearities. Figure 10.7 shows the income-by-education data again with the linear fit, a quadratic (parabolic; second degree polynomial) fit, and a cubic (third degree polynomial—x^3) fit. The quadratic fit is better than the linear fit, but it fails to capture the higher incomes at the lowest levels of schooling, and it overestimates incomes at the very highest levels of education. The cubic fit appears to be the best. At the same time, however, the interpretation of the cubic model, which involves three terms for education—x, x^2, and x^3—is difficult. Consequently, the quadratic model is probably preferable, especially given that the sample size at the lowest levels of schooling is quite small, thus making the sample means for income highly variable at these levels.

10.4 Conclusions

In this chapter, we began by discussing the requirements that must be met in order to establish that a relationship between two variables is causal. As we saw, experimental methods are the gold standard for establishing causality, because randomization of sample members to treatment and control groups allows us to rule out spuriousness, that is, alternative explanations for the relationship between the variables of interest. However, social science research generally involves treatments that cannot be randomly assigned to respondents. Thus, social scientists tend to turn to multiple regression methods. We showed the basic extension of the simple regression model to handle multiple independent variables. We then extended the multiple regression model to handle noncontinuous independent variables, statistical interactions between independent variables, and nonlinear relationships between x and y variables. Together, these extensions of the model make the multiple regression model highly flexible and therefore extremely useful for statistical analysis of social science data. Indeed, because of its flexibility, the linear regression model, and further extensions of it, are widely used in research.

10.5 Items for Review

- Causality rules
- Spuriousness
- Counterfactual model
- True experiment
- Treatment
- Control
- Placebo
- Randomization
- Statistical control
- Interpreting coefficients in multiple regression
- F test in multiple regression
- t tests on parameters
- Total, direct, and indirect effects
- Suppressor effects
- Dummy variables
- Statistical interaction
- Capturing nonlinearity in regression

10.6 Homework

1. The following table presents three models for the relationship between education and income using GSS data only up to 2006. For all three, plot the prediction curve.

Variable	Coefficient		
	Model 1	Model 2	Model 3
Intercept	−2.99***	8.50***	24.95***
Education	4.10***	2.11***	−3.53***
Education^2		.08***	.63***
Education^3			−.016***

Regression of income on polynomials for education, 1972–2006 GSS data, n = 26,228. (*** $p < .001$)

2. Below is a table with results of a single regression model predicting income. The model contains a statistical interaction between sex and education. Plot the implied regression lines across education for men and women.

Variable	Coefficient
Intercept	−28.34***
Male	23.43**
Education	5.62***
Male*Educ.	−1.16*

Regression of income on sex and education, 2000 GSS data, n = 1,667 (* $p <$ .05, ** $p < .01$, *** $p < .001$)

3. Below is a regression model predicting income. Birth cohort is constructed as year of survey minus age of respondent. Male, married, and lives in south are dummy variables with references as female, not married, and lives in other regions. Education is in years of schooling, and health is measured so that a higher score indicates better health. Interpret the results.

Variable	Coefficient
Intercept	−27.96***
Age	.09***
Birth cohort	.04*
Male	5.3***
Education	3.6***
Married	21.7***
Lives in South	−.74
Health	5.27***
R^2	.29

Regression of income on selected covariates, 1972–2006 GSS data, n = 26,228 (* $p < .05$, ** $p < .01$, *** $p < .001$)

4. The GSS asks a question each year about political party affiliation, with 0 being a "strong Democrat" and 6 being a "strong Republican." Values in between reflect less extremity, with 3 being moderate, independent, or unsure (as per my coding of response categories). I estimated a series of regression models as shown in the table below, for persons ages 30–64, from 1972 to 2010 to examine the pattern of party preference over time for males and females. Plot the prediction curves for each gender across year for each model and interpret. How different is the interpretation for Model 1 vs. Model 4?

Variable	Model 1	Model 2	Model 3	Model4
Intercept	1.21***	−6.04***	27.09*	29.92**
	(.13)	(1.15)	(11.01)	(11.01)
Male	.34***	.34***	.34***	−.967***
	(.03)	(.03)	(.03)	(.25)
Year	.015***	.174***	−.923*	−.997**
	(.001)	(.023)	(.36)	(.36)
$Year^2$		−.00086***	.0111**	.0118**
		(.0001)	(.004)	(.004)
$Year^3$			−.000044**	−.000046**
			(.00001)	(.00001)
Male*Year				.0141***
				(.003)
R^2	.012	.014	.015	.016

Regression of party affiliation on selected covariates, 1972–2010 GSS data, ages 30–64, n = 20,409 (* $p < .05$, ** $p < .01$, *** $p < .001$)

5. Black and white differences in health are widely studied in social science research. It is well known that whites report better health on average than blacks, and whites have longer life expectancies, as well. A key question is why. Below is a path diagram with standardized regression coefficients for direct and indirect paths from race to health. Based on the diagram, what is the total effect of race on health? What proportion of the total effect is explained by education and income differences between blacks and whites? (hint: there are THREE indirect paths from black to health).

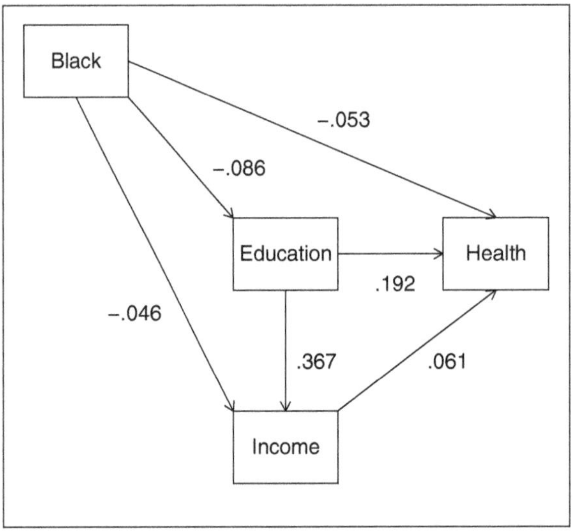

Path diagram for direct and indirect effects of race on health (1972–2010 GSS data, ages 30–64, n = 14,149, persons of "other" races excluded).

6. Suppose I am interested in determining why some people are Republicans, while others are Democrats. So, I regress the party affiliation variable described above on a number of demographic and social variables and obtain the results shown in the table below. I have a friend with the following characteristics: He was born in 1971, he is white and lived in the south when he was 16. His parents' education level is 20 years, while his own education level is 19. He is married, currently lives in the south, he claims excellent health, and he makes $150,000 per year in income. Compute his predicted score on the party affiliation variable and speculate about his affiliation.
   Note: cohort is computed as year (as a three digit number, with 2000 being 100) minus age (so, if cohort = 71, current age in 2013 would be 42). Male, Black, Other, South at 16, Married, and South are all dummy variables. Health is rated on a 4 point scale coded as: 0 = Poor, 1 = Fair, 2 = Good, 3 = Excellent. Income is in $1,000 units. The outcome variable, as described above, ranges from 0 (strong Democrat) to 6 (strong Republican). The central value on the scale—3—is not affiliated with either party.

Variable	Coefficient
Intercept	.72(.20)***
Cohort	.014(.002)***
Age	.014(.002)***
Male	.200(.03)***
Black	−1.58(.05)***
Other	−.53(.07)***
South at 16	−.09(.05)#
Parents' Education	.04(.005)***
South	.17(.05)***
Married	.28(.03)***
Education	−.006(.006)
Health	.07(.02)**
Income	.002(.001)*
R^2	.11

Regression of party affiliation on selected covariates, 1972–2010 GSS data, ages 30–64, n = 14,835 (* $p < .05$, ** $p < .01$, *** $p < .001$)

Chapter 11
Presenting Results of Statistical Analysis

This chapter is intended to be a practical guide to help with the construction of tables and figures and with the general presentation of results of statistical analysis in a research paper. Constructing tables and figures well and writing a results section so that it appears to make a coherent point—and does not wander—is as important as constructing a solid research question and conducting analyses correctly to answer it. That is, if the reporting of the results does a poor job of telling a story that can answer the research question, it is ultimately pointless to have developed a good literature review and research question. For that matter, the statistical analyses may have been performed extremely well, but if the results are not displayed in a way that is easy to see and understand, the analyses have been a waste of time.

The data analysis process should, if it is done well (i.e., the data have been thoroughly investigated), yield far more results than one can possibly report in a single paper. Part of the art of writing a good paper is determining what should be reported (and in what order) to make a good story. I recommend that, before one begins to actually write the results section, one should (1) decide what results will be reported, (2) make tables and figures and put them in the order to be discussed, and (3) plan to write the results from the tables and figures.

11.1 Making Tables

Tables are the key way that results of statistical analyses are presented. A table may present descriptive statistical information, or it may report results of inferential analyses and statistical tests. In general, tables should be easy to read, should not contain unnecessary information, but should contain all information relevant to understanding the analyses, and they should be able to stand alone from the text. That is, one should not have to reference the text of a paper to understand a table. With these ideas in mind, there are a number of important rules for making tables. Some of them are guidelines more than rules; different journals, researchers, and disciplines have different styles.

S.M. Lynch, *Using Statistics in Social Research: A Concise Approach*,

DOI 10.1007/978-1-4614-8573-5_11, © Springer Science+Business Media New York 2013

	Mean or Percent (s.d.) [range]	
Variable	**Male (n = 888)**	**Female (n = 1148)**
Age	47.8(17.0)[18,89]	48.1(18.1)[18,89]
Nonwhite	22.6 %	25.3 %
Education (yrs.)	13.4(3.3)[0,20]	13.5(3.1)[0,20]
Income ($1000)	31.0(31.1)[0,119.6]	24.4(27.5)[0,119.6]

Note: Income statistics include persons who have income of 0.

Table 11.1. Descriptive statistics for variables used in analyses by sex (2010 GSS).

First and foremost, *do not* simply cut and paste raw output from a statistics software package. Raw output from a statistics package does not constitute a table and is insufficient as the body of the table. The material is not easy to read, and there is usually more material reported than is necessary for discussion. Furthermore, the material is usually not presented in the most readable format.

Second, tables must have titles, and the title must fully describe what the table contains. It is insufficient to have a title like: "Table 11.1. Results." Instead, the title should be detailed: "Table 11.1. Results of Regression Analyses of Income on Education and Background Variables (NHANES sample, n = 3,076)." If there is too much information to put in the title, add a footnote. For example, in the previous title, I may wish to footnote the table in order to define "NHANES."

Third, beyond any additional information that cannot be fit into the title, most tables of results have footnotes reporting symbols for levels of statistical significance. The usual convention in social science research is: "# $p < .10$, * $p < .05$, ** $p < .01$, *** $p < .001$."

Fourth, tables generally consist of columns. The first column is almost always a list of the variables for which descriptive statistics are being reported or for which regression coefficients are being reported. Subsequent columns are used for (1) the regression coefficients or descriptive statistics, and (2) results from additional models (if more than one model is being reported in the table) or results for different groups that were modeled (e.g., races, sexes, etc.).

Fifth, variable names should be recognizable. Do not use variable names like "v2103," or whatever variable name existed in the data set; instead, use names like "Male" or "Nonwhite." In fact, for dummy variables, name them after what they indicate, as with "male;" do not use "sex," because it is then necessary to define in a footnote what value is 1 vs. 0.

Sixth, usually, the regression coefficients (or means or percents, if the table is reporting descriptive statistics), are reported first, followed by the standard errors of the coefficients in parentheses (or standard deviations, if the table is reporting descriptive statistics).

Finally, given that virtually all tables have footnotes, I recommend the following format for any table: (1) Title in bold, (2) one or two horizontal lines, (3) main

	Male (n = 807)		Female (n = 994)	
Variable	Model 1	Model 2	Model 1	Model 2
Intercept	−20.95*** (5.22)	−20.43*** (5.72)	−27.96*** (4.58)	−34.94*** (5.14)
Age	.09 (.06)	.09 (.06)	.08# (.05)	.08# (.05)
Nonwhite	−5.82* (2.42)	−7.96 (9.79)	−8.83*** (1.83)	13.82# (7.91)
Education	3.85*** (.31)	3.81*** (.35)	4.00*** (.27)	4.51*** (.32)
Nonw*Educ		.17 (.74)		−1.70** (.58)
R^2	.18	.18	.22	.22

# $p < .1$, * $p < .05$, ** $p < .01$, *** $p < .001$
[a] The sample was reduced to individuals who had non-zero earnings.
Table 11.2. Results of multiple regression analyses of income on race and education by sex (2010 GSS)[a].

body of the table, (4) one horizontal line, (5) the p-value symbol note, (6) additional footnotes. The title of a table may, alternatively, be placed at the bottom of the table as done throughout this book.

Tables 11.1 and 11.2 illustrate table construction. Table 11.1 reports descriptive statistics; Table 11.2 reports the results of a set of regression models. These tables demonstrate the rules/guidelines shown above. The title of each table describes what the table contains, including reference to the sample. Here, the sample name (GSS) is an abbreviation and should probably be spelled-out in a footnote, but in most journals that would publish a paper using these data, the abbreviation would be known by readers. Both tables contain a considerable amount of information. Table 11.1 provides detailed summary statistics by two sex groups in the sample for four variables.

Table 11.2 provides regression coefficients for multiple regression models predicting income conducted for each sex. In one model for each sex (Model 2), an interaction between race and education is included; in the other model (Model 1) only main effects are included.

11.2 Making Figures

In addition to tables, figures are an important means for presenting interrelationships between concepts in a theory (a "conceptual plot") and presenting results of analyses or showing patterns, trends, and relationships between variables in data. The most common data-based figures used in social science papers include histograms, bar charts, scatterplots, and line plots. Histograms display the frequency distribution of a variable (see earlier chapters). Bar charts represent means or totals of some variable for different groups. The x-axis of these plots references the different groups, while the y-axis (the height of the bars) is the mean or total of the variable of interest. Scatterplots are used to plot one continuous variable against another so that any relationship between them can be seen. Finally, line plots are often used to show trends in variables across time. They are also often used after a regression analysis to show model-predicted patterns/relationships between variables. Each of these types of figures, along with some others, has been used throughout this book.

A good figure is one that conveys considerable information very simply and does not mislead a viewer. To that end, there are a number of rules for making figures. Some of the rules are the same as for tables, including that a figure must stand alone without relying on reference to the text. There are a number of additional rules for constructing good figures.

First, as with tables, plots that are produced in statistics packages using default settings do not constitute legitimate figures. They will usually violate other rules that we will discuss. Instead, the data for figures will generally need to be entered into Excel or some other package that is specifically geared for producing figures.

Second, figures must have titles, and the title must fully describe what the figure contains. If plotting one variable against another, the general rule for phrasing is that Y—the variable on the y-axis—is plotted against X.

Third, variable names should be recognizable in the figure, just like they should be in a table. Don't use variable names like "v2103," or whatever variable name existed in the data set; instead, use names like "Male" or "Nonwhite."

Fourth, label all axes with appropriate labels. Do not leave this to the default settings of the software package.

Fifth, choose the scale of the axes appropriately. Do not truncate the domain or range of the data. The origin (0,0) should almost always be included to prevent distortion of the relationship between the variables in the plot. Do not leave the choice of domain or range of axes to the default settings of the software package.

Finally, do not make the figure more complex or fancier than it needs to be to display the material. For example, bar graphs do not necessarily need to have colored bars, nor do the bars need to be three-dimensional. In scatterplots, plotting characters do not generally need to be more complex symbols than dots or circles. In brief, a good figure is one that is not confusing, not one that has a pretty or sophisticated appearance.

Figures 11.1 through 11.3 highlight some of these rules. All three figures use the same data, but they appear very different and have different levels of readability.

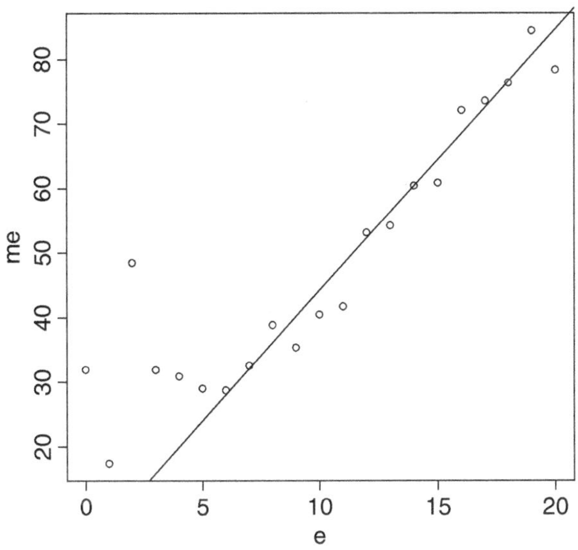

Fig. 11.1. Income and schooling

Figure 11.1 shows the relationship between education and income by plotting mean income for persons at each level of education—measured in years—in the GSS. The plot is the default plot obtained using my preferred software (R). This figure violates almost every rule for making figures. The title of the figure is insufficient to understand what the figure shows. What are the circles in the figure, and what is the line? Where did the data come from? The axes are labeled poorly: what are "e" and "me"? Finally, the image itself provides a somewhat misleading impression of the strength of the education-income relationship. Notice that the y-axis in the plot starts at 20, rather than 0. Thus, the y dimension of the plot is compressed, making the slope of the line in the figure appear quite steep.

Figure 11.2 remedies almost all of the problems that were present in Fig. 11.1. The title is much more detailed. The axis labels have been replaced with words, so that it is very clear that the x-axis refers to years of schooling, ranging from 0 to 20, while the y-axis refers to mean income. The label clarifies that the units are thousands of dollars. Importantly, the axis range now starts at 0, and the axis has been expanded upward to 100 (again, in thousands). Obviously, we could increase the upper limit further, but all of the data points seem to fall pretty evenly between 0 and 100, and 100 is a reasonable upper limit for income in the US population. Very few earn more than this. Finally, the inclusion of a legend clarifies that the observed means in the data are represented by the circles, while the line is a predicted regression line, showing the smoothed expected relationship between years of schooling and earnings. Although not shown in the figure, each year of schooling appears to be associated with a $4,000 increase in income.

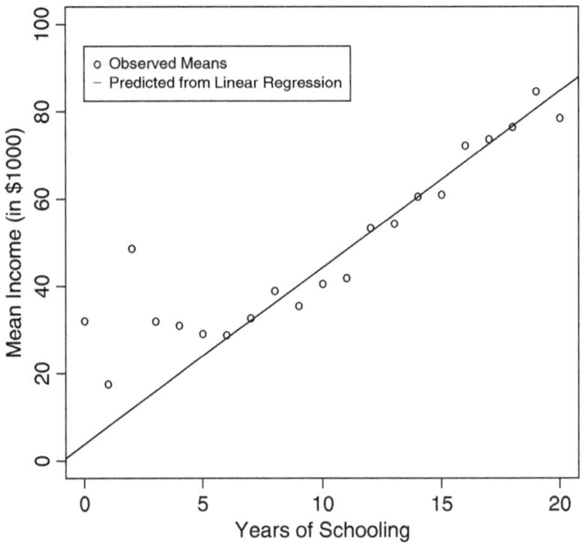

Fig. 11.2. Observed and predicted mean income by years of education (n = 12,924, 1972–2006 GSS data for persons ages 30–54; income in 2006 dollars).

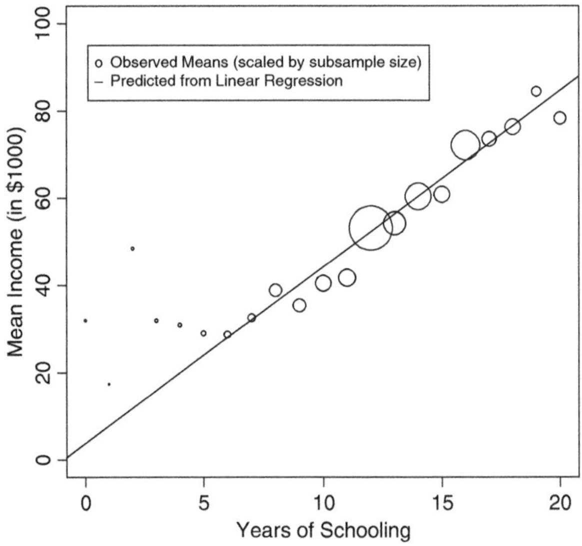

Fig. 11.3. Observed and predicted mean income by years of education (n = 12,924, 1972–2006 GSS data for persons ages 30–54; income in 2006 dollars).

One problem that remains in the figure is that the regression line seems to fit the observed means somewhat poorly, especially at the lowest levels of education. Figure 11.3 clarifies why: There are very few individuals in the sample at these levels of education, and so the regression results are driven primarily by those with more schooling. The final figure demonstrates this by using circles at each level of schooling that are sized based on the number of individuals in the sample at that level of schooling.

11.3 Writing the Results Section

Once tables and figures have been created and placed in the order in which they are to be discussed, writing the results section of a paper is relatively simple. Remember that tables and figures must stand alone; that is, they must be able to be viewed without reference to the text. Similarly, the results section of the paper must be readable without reference to the tables and figures. This does not mean that one cannot refer to the tables and figures in the text—indeed, they should be referenced. It simply means that a reader must be able to understand the results section without looking at the tables and figures. To some extent, then, the results section and the tables and figures may seem somewhat redundant. However, the tables and figures are provided to supplement the text and give a quick, visual summary of the writing in the results section.

I recommend printing each table and figure on a separate page, placing them in order, and then writing the results section as if you were explaining the tables and figures to someone verbally, walking through the table and figure results one step at a time, following the storyline. Every single numerical result that is presented in the tables and figures need not be tediously addressed. Instead, focus on the main highlights. For example, in presenting the results from Table 11.2 (the regression model results), the intercepts in each model do not need to be discussed, nor do the standard errors. Instead, something like the following is sufficient:

> The regression model results presented in Table 11.2 show that older persons uniformly tend to earn slightly more than younger persons, although the results in each regression model are either not statistically significant or only marginally so. For example, among men, Models 1 and 2 show that each year of age produces an increase of $90 in earnings, although the result is not statistically significant. Among women, each year of age produces an expected increase of $80 in earnings, but this result is only marginally significant ($p < .1$). For both sexes, Model 1 results show that each year of schooling is worth approximately $4000 more in earnings ($3850 for men; $4000 for women). Model 1 also shows that nonwhites earn less than whites. The race difference in earnings is greater for women than for men, however. Nonwhite men earn about $6000 less than white men, while nonwhite women earn about $9000 less than white women.
>
> The results of Model 2, in which an interaction between race and education is included, show that the returns to education for whites and nonwhites vary across sex. For men, the Model 2 results are very similar to the Model 1 results. The interaction

between race and education is only .17 and is not statistically significant. For women, the Model 2 results are very different than the Model 1 results. In Model 2, nonwhite women with no schooling earn *more* than white women with no schooling, but that advantage erodes with more schooling. Given the drastic change in coefficients from Model 1 to Model 2, some predicted values for earnings provide a clearer picture. For men at age 45 (age is constant for all predictions), whites with 12 years of schooling earn approximately $30,000 on average. The average is roughly double for those with 20 years of schooling. For nonwhite men, average earnings are about $21,000 at 12 years of schooling (a difference of $9000 with white men with the same level of schooling), but about $55,000 at 20 years of schooling—a difference of only $5000 with white men with the same level schooling. Thus, each year of schooling nets slightly more for a black male than a white male.

For women, this pattern is reversed. White women with 12 years of schooling earn approximately $23,000 on average, and that average more than doubles to $59,000 at 20 years of schooling—almost as much as men with 20 years of schooling. For nonwhite women, average income is $37,000 for those with 12 years of schooling, but the average only increases another $2000 for those with 20 years of schooling.

Overall, the results indicate that nonwhite men gain slightly more per year of schooling than white men (albeit not a statistically significant amount), and income increases substantially with education. In contrast, while white education at low levels of education earn far less than white males, nonwhite women earn considerably more than nonwhite males at low levels of schooling. However, white women gain far more from schooling than nonwhite women, approaching the earnings of their male counterparts, while nonwhite women appear to gain little from additional schooling, thus earning substantially less than nonwhite males.

Based on how I wrote this description of the table, it is apparent that my research question is/was focused on whether race differences in the return to education varied by sex (a three-way interaction). If my focus had been simply on education's effect by sex, I would have focused on discussing the fact that the Model 1 results show that men and women appear to receive roughly the same returns to schooling (coefficient of 3.85 vs. 4.00), and I would have excluded Models 2.

Notice how, in this write-up of the results of the table, I did not discuss every number in the table. Results sections of papers are often inherently boring to read, and so you should focus on the key results to keep the reader interested and to keep the focus on the story you are telling. Also notice how I did not make any value judgments about the results, and I did not draw any implications regarding my theory and research question: This discussion is for the "Discussion" section of the paper. The results section simply presents the results.

11.4 Writing the Discussion Section

The Discussion section of the paper summarizes the results further and discusses how the results answer the research question, and more broadly, what the implications of the results are for the larger theory from which the research question was drawn. In this particular example, I may wish to say something like:

> Stratification research has long focused on both racial and gender differences in earnings (citations), as well as on the importance of education to earnings (citations), and to a lesser extent the differential returns to education for men versus women and whites versus persons of other races (citations). However, little research has simultaneously considered that gender returns to education may themselves vary by race. Yet, such a consideration is theoretically important (why?; citations)...In this study, we found that there is a substantial difference by sex in how education affects earnings for whites and nonwhites. While white men at low levels of schooling earn considerably more than nonwhite men at comparable levels, the earnings gap shrinks substantially across levels of schooling. The pattern is much different for women. Nonwhite women earn more than white women at lower levels of schooling, but their wages are relatively flat across education. White women, in contrast see large returns to each additional year of schooling so that they earn nearly as much as white men at the highest levels of schooling. All in all, the results support a "double jeopardy" hypothesis: that being both female and a member of a racial minority present a double threat to earnings that education simply does not overcome. The implications of these findings include that research on discrimination, and stratification research more broadly, should not focus on sex or race alone; instead, research should consider the nexus of these two social groupings as suggested by intersectionality theory...

Notice how, in this excerpt, I have not discussed particular, numerical results like I did in the Results section. Also notice how I placed my findings (and my study more generally) in the context of previous research, and I have drawn implications for subsequent research. (as a side note, research in stratification has already considered these issues; this is simply an example).

The end of the discussion section usually points to limitations of the present research presented in the paper, and it offers suggestions for future study. Sometimes this information is presented in a separate Conclusions section; sometimes the paper ends at the end of the Discussion. In this example—just to provide one of many shortcomings—I might point out that I did not control on employment status other than to exclude persons who reported no income. It is quite possible that differences by sex and race in the proportion engaged in part time work explains some of the findings, and this limitation should be mentioned (if not corrected in the original analyses!)

11.5 Conclusion

Making good tables and figures is crucial to writing a good research paper. In this chapter, we have discussed some basic rules for constructing good tables and figures, as well as writing summaries of them. At this point, we have covered the entire process of conducting basic quantitative research, from developing a research question, to constructing a survey instrument, to obtaining data, to analyzing it using basic statistical methods, to reporting the results. You should therefore be ready to conduct your own quantitative research from start to finish. In the final chapter, I offer some suggestions for additional reading to help flesh out each of the topics we have discussed throughout the book.

Chapter 12
Conclusion

We have covered a lot of ground in this book. We began in Chap. 2 by discussing the process of scientific research, including the philosophical underpinnings of the scientific method. As we discussed, this process begins with the construction of a succinct research question and hypothesis that has the key property of being falsifiable. As we showed, the goal of scientific research is to falsify hypotheses derived from larger explantions for how the world works. We can never prove our hypotheses to be true, and so the best we can do is rule out bad hypotheses. Explanations that generate hypotheses that cannot be rejected after extended testing eventually become recognized as theories, the ultimate status for a scientific explanation of the world.

The process of hypothesis testing requires the collection of data from the real world, often via observation, interviews, and surveys. We discussed the difficulties and goals of data collection in Chap. 3, but we barely scratched the surface. Entire courses are offered on research methodology, including courses only on sampling methods. The books for such courses range from general books on the theory and practice of research (e.g., Babbie 2004; Firebaugh 2008; Lieberson 1987) to books on the technical details of data collection (e.g., Dillman et al. 2009).

In Chap. 4, we began discussing how to summarize data with basic descriptive measures that reflect the center and spread of the distribution of variables. We then discussed the limitations of such measures and their misuse. There are a number of excellent and entertaining books that discuss an array of ways in which statistics can be abused (e.g., Campbell 2004; Hooke 1983; Huff 1993). Such books are not only often fun to read, but they are also very instructive in showing how to be a critical consumer of statistical data and arguments based on them.

We turned our attention in Chap. 5 away from descriptive methods and discussed probability theory. Probability is the foundation on which inferential statistics is based, and understanding probability is crucial to understanding and evaluating statistical arguments. Our discussion of probability, while fairly extensive, was not completely thorough. Other excellent books exist that cover probability theory in much greater depth (e.g., DeGroot and Schervish 2012). While most people think in probabilistic terms—and probabilistic terminology permeates our language

S.M. Lynch, *Using Statistics in Social Research: A Concise Approach*,

DOI 10.1007/978-1-4614-8573-5_12, © Springer Science+Business Media New York 2013

and discussions—people are often bad at formal probabilistic reasoning. We often overestimate or underestimate probabilities and evaluate them incorrectly, reaching poor conclusions. Although we discussed some of the ways in which probabilistic reasoning can be flawed, there are a number of books available that provide entertaining discussions of common fallacies in probabilistic thinking and mathematical thinking more generally (e.g., Bunch 1997; Campbell 2004; Mlodinow 2008; Paulos 2001). There are also several books that show how such fallacious reasoning and analyses have led to real-world policy emphases that do not serve the public interest (e.g., Agin 2006; Pigliucci 2010).

Chapter 6 illustrated one of the most crucial theorems—the Central Limit Theorem—that helps us make the leap from deductive, probabilistic thinking to more inductive, inferential reasoning. That chapter shows the basis for understanding why small, random samples are all that is needed to make valid and precise statements about huge, even infinite, populations. We also discussed how we can use inferential methods to evaluate statistical hypotheses. The most important idea, one that has been repeated again and again in subsequent chapters, is that classical statistical hypothesis testing involves evaluating the probability of observing the sample data we obtained if the hypothesis we are evaluating were true. If that probability is small, we then reject the hypothesis (see DeGroot and Schervish 2012). This approach to statistical hypothesis testing seems backwards to many and I believe is what makes learning statistics difficult. In all fairness, one should become familiar with the criticism of the approach to statistics involving p-values that we discussed in this book (see Ziliak and McCloskey 2008). Furthermore, there is a competing paradigm of statistics—Bayesian statistics—that more directly assesses the probability that hypotheses are true and corresponds better, arguably, to how people actually think. Numerous, more advanced statistical courses and books are available to learn more about this approach to statistics (see Lynch 2007).

Chapters 7 through 9 extended the basic precepts of statistical inference and hypothesis testing to different types of data, that is, data measured at different levels. While we have covered some basic methods for assessing relationships between different combinations of nominal, ordinal, and continuous variables, we have limited our discussion to some key, basic approaches. Entire courses can certainly be taken on methods for each combination of different types of measures. Such courses are usually taken after a basic course on multiple regression modeling, the basics of which we discussed in Chap. 10.

Chapter 10 began with a discussion of the notion of causality and the difficulties with establishing that a relationship between two variables is a causal one. As we discussed, experimental methods are usually viewed as the gold standard for establishing that relationships are causal, but in most social science research, experimental methods are impossible to employ. Multiple regression serves as a key method for establishing the relationship between two or more variables while simultaneously controlling out relationships that potentially explain the relationship of interest (see Morgan and Winship 2007, for extensive discussion of the state of causal modeling in social science). Regression modeling is a fundamental approach to handling "multivariate" data, that is, multiple variables, and we barely scratched the surface in introducing it. The next course one usually takes in statistics is a detailed course on multiple regression modeling that spells out the statistical assumptions that underlie the model, the consequences of violating those assumptions, and methods for compensating for them. There are a number of excellent books available that elaborate multiple regression methods and show how it can be extended in numerous ways to handle different data contexts, like time series data and panel data (e.g., Fox 2008; Gujarati and Porter 2009).

Finally, in Chap. 11, we discussed how to present the results of statistical analyses. Learning how to present results of statistical analyses via tables and figures is easily as important as learning how to conduct the analysis itself. Poor presentation can make the results unintelligible at best and misleading at worst. From a practical perspective, if you have gone to the trouble to conduct good analyses, you should certainly want to convey what you've found in a way that people can follow! Chap. 11 provided very basic rules for doing so, but as with the material in other chapters, there are many books on the topic that should be explored (e.g., Cleveland 1993).

In sum, we have covered a lot of ground in this book, but there is far far more out there to be learned. I hope that you have found the material we covered to be interesting, and even fun, and I wish you luck in conducting your own analyses!

Appendix A
Statistical Tables

A.1 The Z (Standard Normal) Distribution

The normal distribution—the "bell curve"—is a continuous, symmetric distribution that is peaked in the center and tapers away from the center. The distribution represents how many natural phenomena are distributed: most observations in a population cluster around the mean, and the more extreme an observation is relative to the mean, the rarer such an observation is. The distribution is important not only as a representation of the distribution of many items in the natural world, but also as the key distribution used for statistical inference under the Central Limit Theorem discussed in Chap. 6. According to the theorem, sample statistics—in particular sample means—obtained (theoretically) under repeated random sampling follow a normal distribution centered over the true population mean and with a standard deviation equal to the population standard deviation divided by the square root of the size of the (repeated) samples. Thus, under the theorem, we can quantify our uncertainty about using a sample statistic as an estimate of a population parameter.

Figure A.1 shows the standard normal (z) distribution and the quantities presented in the z table. Due to symmetry, all desired regions under the curve can be obtained from the table. Equally important, area of regions under *any* normal distribution can be obtained by transforming the original data to have a mean of 0 and standard deviation of 1 as follows:

$$z = \frac{x - \mu}{\sigma}, \tag{A.1}$$

where x is the point of interest under the original distribution, μ is the population mean, and σ is the population standard deviation. For hypothesis testing and confidence interval construction, σ should be the *standard deviation of the sampling distribution*, which is computed as $\sigma/\sqrt{n}$ (called the standard error) (Table A.1).

S.M. Lynch, *Using Statistics in Social Research: A Concise Approach*,
DOI 10.1007/978-1-4614-8573-5, © Springer Science+Business Media New York 2013

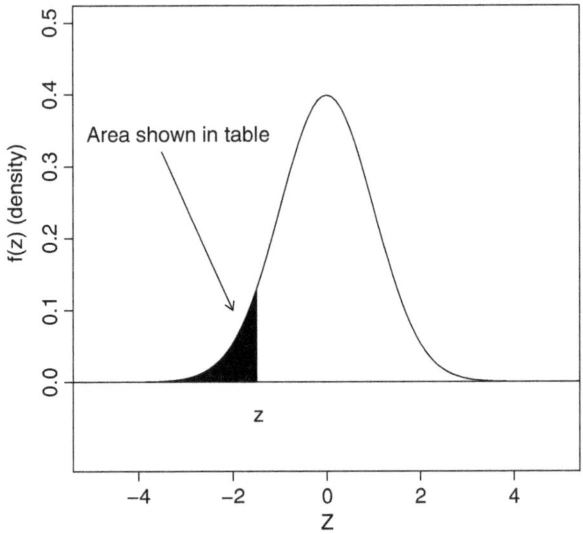

Fig. A.1. The standard normal distribution

Z	Third digit of Z -0.09	-0.08	-0.07	-0.06	-0.05	-0.04	-0.03	-0.02	-0.01	-0.00
-3.2	0.001	0.001	0.001	0.001	0.001	0.001	0.001	0.001	0.001	0.001
-3.1	0.001	0.001	0.001	0.001	0.001	0.001	0.001	0.001	0.001	0.001
-3.0	0.001	0.001	0.001	0.001	0.001	0.001	0.001	0.001	0.001	0.001
-2.9	0.001	0.001	0.001	0.002	0.002	0.002	0.002	0.002	0.002	0.002
-2.8	0.002	0.002	0.002	0.002	0.002	0.002	0.002	0.002	0.002	0.003
-2.7	0.003	0.003	0.003	0.003	0.003	0.003	0.003	0.003	0.003	0.003
-2.6	0.004	0.004	0.004	0.004	0.004	0.004	0.004	0.004	0.005	0.005
-2.5	0.005	0.005	0.005	0.005	0.005	0.006	0.006	0.006	0.006	0.006
-2.4	0.006	0.007	0.007	0.007	0.007	0.007	0.008	0.008	0.008	0.008
-2.3	0.008	0.009	0.009	0.009	0.009	0.010	0.010	0.010	0.010	0.011
-2.2	0.011	0.011	0.012	0.012	0.012	0.013	0.013	0.013	0.014	0.014
-2.1	0.014	0.015	0.015	0.015	0.016	0.016	0.017	0.017	0.017	0.018
-2.0	0.018	0.019	0.019	0.020	0.020	0.021	0.021	0.022	0.022	0.023
-1.9	0.023	0.024	0.024	0.025	0.026	0.026	0.027	0.027	0.028	0.029
-1.8	0.029	0.030	0.031	0.031	0.032	0.033	0.034	0.034	0.035	0.036
-1.7	0.037	0.038	0.038	0.039	0.040	0.041	0.042	0.043	0.044	0.045
-1.6	0.046	0.046	0.047	0.048	0.049	0.051	0.052	0.053	0.054	0.055
-1.5	0.056	0.057	0.058	0.059	0.061	0.062	0.063	0.064	0.066	0.067
-1.4	0.068	0.069	0.071	0.072	0.074	0.075	0.076	0.078	0.079	0.081
-1.3	0.082	0.084	0.085	0.087	0.089	0.090	0.092	0.093	0.095	0.097
-1.2	0.099	0.100	0.102	0.104	0.106	0.107	0.109	0.111	0.113	0.115
-1.1	0.117	0.119	0.121	0.123	0.125	0.127	0.129	0.131	0.133	0.136
-1.0	0.138	0.140	0.142	0.145	0.147	0.149	0.152	0.154	0.156	0.159

(Continued)

-0.9	0.161	0.164	0.166	0.169	0.171	0.174	0.176	0.179	0.181	0.184
-0.8	0.187	0.189	0.192	0.195	0.198	0.200	0.203	0.206	0.209	0.212
-0.7	0.215	0.218	0.221	0.224	0.227	0.230	0.233	0.236	0.239	0.242
-0.6	0.245	0.248	0.251	0.255	0.258	0.261	0.264	0.268	0.271	0.274
-0.5	0.278	0.281	0.284	0.288	0.291	0.295	0.298	0.302	0.305	0.309
-0.4	0.312	0.316	0.319	0.323	0.326	0.330	0.334	0.337	0.341	0.345
-0.3	0.348	0.352	0.356	0.359	0.363	0.367	0.371	0.374	0.378	0.382
-0.2	0.386	0.390	0.394	0.397	0.401	0.405	0.409	0.413	0.417	0.421
-0.1	0.425	0.429	0.433	0.436	0.440	0.444	0.448	0.452	0.456	0.460
-0.0	0.464	0.468	0.472	0.476	0.480	0.484	0.488	0.492	0.496	0.500

Table A.1. Table shows LEFT TAIL probabilities—area from $-\infty$ up to Z

A.2 The *t* Distribution

The t distribution is similar in shape to the normal distribution, but has heavier tails. The heavier tails stem from uncertainty in using s as an estimate of σ in a normal distribution problem. Uncertainty about σ declines rapidly as n increases, and so, as the "degrees of freedom" ($df = n - 1$) increase, the t distribution becomes increasingly like the normal distribution. Indeed, when $n > 120$, one can simply use the z table, because t and z values coincide at least two decimal places.

The table shows the right-hand "critical value" of t needed to obtain a given one-tailed probability (α) in the right-hand tail (Table A.2). Figure A.2 shows the t distribution with 1 df.

	ONE-TAILED α						
df	**.25**	**.20**	**.15**	**0.1**	**0.05**	**0.025**	**0.005**
1	1.00	1.38	1.96	3.08	6.31	12.71	63.66
2	0.82	1.06	1.39	1.89	2.92	4.30	9.92
3	0.76	0.98	1.25	1.64	2.35	3.18	5.84
4	0.74	0.94	1.19	1.53	2.13	2.78	4.60
5	0.73	0.92	1.16	1.48	2.02	2.57	4.03
6	0.72	0.91	1.13	1.44	1.94	2.45	3.71
7	0.71	0.90	1.12	1.41	1.89	2.36	3.50
8	0.71	0.89	1.11	1.40	1.86	2.31	3.36
9	0.70	0.88	1.10	1.38	1.83	2.26	3.25
10	0.70	0.88	1.09	1.37	1.81	2.23	3.17
11	0.70	0.88	1.09	1.36	1.80	2.20	3.11
12	0.70	0.87	1.08	1.36	1.78	2.18	3.05
13	0.69	0.87	1.08	1.35	1.77	2.16	3.01
14	0.69	0.87	1.08	1.35	1.76	2.14	2.98
15	0.69	0.87	1.07	1.34	1.75	2.13	2.95
16	0.69	0.86	1.07	1.34	1.75	2.12	2.92
17	0.69	0.86	1.07	1.33	1.74	2.11	2.90
18	0.69	0.86	1.07	1.33	1.73	2.10	2.88
19	0.69	0.86	1.07	1.33	1.73	2.09	2.86

(Continued)

20	0.69	0.86	1.06	1.33	1.72	2.09	2.85
21	0.69	0.86	1.06	1.32	1.72	2.08	2.83
22	0.69	0.86	1.06	1.32	1.72	2.07	2.82
23	0.69	0.86	1.06	1.32	1.71	2.07	2.81
24	0.68	0.86	1.06	1.32	1.71	2.06	2.80
25	0.68	0.86	1.06	1.32	1.71	2.06	2.79
26	0.68	0.86	1.06	1.31	1.71	2.06	2.78
27	0.68	0.86	1.06	1.31	1.70	2.05	2.77
28	0.68	0.85	1.06	1.31	1.70	2.05	2.76
29	0.68	0.85	1.06	1.31	1.70	2.05	2.76
30	0.68	0.85	1.05	1.31	1.70	2.04	2.75
40	0.68	0.85	1.05	1.30	1.68	2.02	2.70
50	0.68	0.85	1.05	1.30	1.68	2.01	2.68
60	0.68	0.85	1.05	1.30	1.67	2.00	2.66
70	0.68	0.85	1.04	1.29	1.67	1.99	2.65
80	0.68	0.85	1.04	1.29	1.66	1.99	2.64
90	0.68	0.85	1.04	1.29	1.66	1.99	2.63
100	0.68	0.85	1.04	1.29	1.66	1.98	2.63
110	0.68	0.84	1.04	1.29	1.66	1.98	2.62
120	0.68	0.84	1.04	1.29	1.66	1.98	2.62
∞	0.67	0.84	1.04	1.28	1.64	1.96	2.58

Table A.2. Table shows ONE TAIL t statistics (t for right tail α)

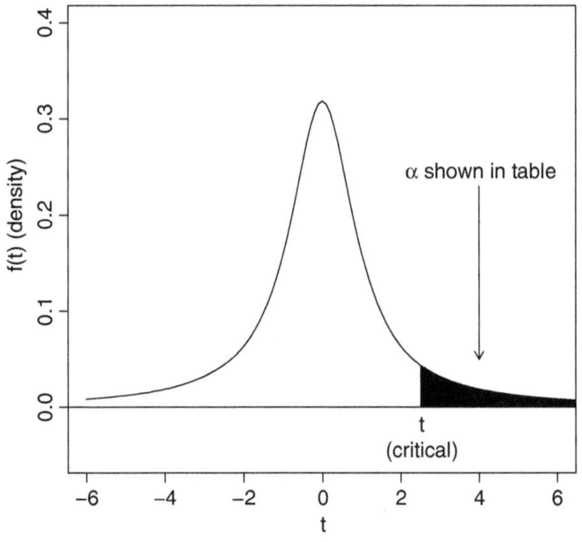

Fig. A.2. The t distribution with 1 degree of freedom

A.3 The Chi-Square (χ^2) Distribution

The chi-square (χ^2) distribution is a distribution for the sum of squared normally-distributed random variables. Because it is a distribution for a sum of squares, the distribution is inherently non-negative. Similar to the t distribution, the χ^2 distribution has a degree of freedom parameter, which corresponds to the number, k, of squared normal variables that have been summed to produce the sum of squares.

Chi-square test statistics are generally tested using one-tail probabilities. Figure A.3 shows a chi-square distribution with 5 df, and the tail probability (α) shown in the table corresponding to the critical value of χ^2 needed to obtain area α (Table A.3). As the χ^2 d.f. increase, the distribution becomes more symmetric and looks more and more like a normal distribution. However, when the d.f. is small, the distribution has a strong right skew because of the left boundary at 0 (Fig. A.4).

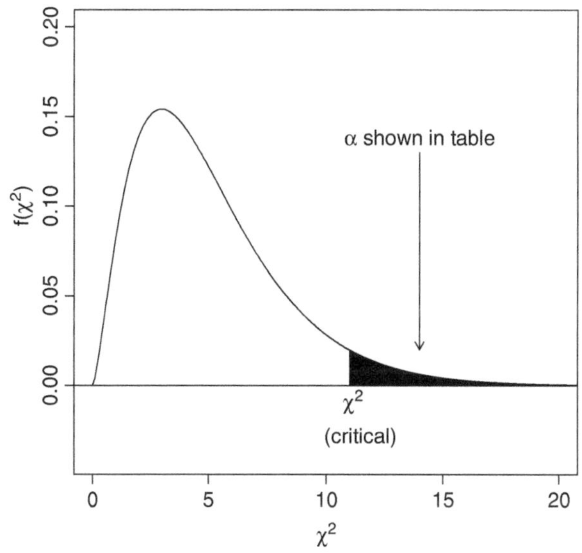

Fig. A.3. The χ^2 distribution with 5 degrees of freedom

	Right tail α						
df	**.25**	**.20**	**.15**	**0.1**	**0.05**	**0.01**	**0.001**
1	1.32	1.64	2.07	2.71	3.84	6.63	10.83
2	2.77	3.22	3.79	4.61	5.99	9.21	13.82
3	4.11	4.64	5.32	6.25	7.81	11.34	16.27
4	5.39	5.99	6.74	7.78	9.49	13.28	18.47
5	6.63	7.29	8.12	9.24	11.07	15.09	20.52
6	7.84	8.56	9.45	10.64	12.59	16.81	22.46
7	9.04	9.80	10.75	12.02	14.07	18.48	24.32
8	10.22	11.03	12.03	13.36	15.51	20.09	26.12

(Continued)

9	11.39	12.24	13.29	14.68	16.92	21.67	27.88
10	12.55	13.44	14.53	15.99	18.31	23.21	29.59
11	13.70	14.63	15.77	17.28	19.68	24.72	31.26
12	14.85	15.81	16.99	18.55	21.03	26.22	32.91
13	15.98	16.98	18.20	19.81	22.36	27.69	34.53
14	17.12	18.15	19.41	21.06	23.68	29.14	36.12
15	18.25	19.31	20.60	22.31	25.00	30.58	37.70
16	19.37	20.47	21.79	23.54	26.30	32.00	39.25
17	20.49	21.61	22.98	24.77	27.59	33.41	40.79
18	21.60	22.76	24.16	25.99	28.87	34.81	42.31
19	22.72	23.90	25.33	27.20	30.14	36.19	43.82
20	23.83	25.04	26.50	28.41	31.41	37.57	45.31
21	24.93	26.17	27.66	29.62	32.67	38.93	46.80
22	26.04	27.30	28.82	30.81	33.92	40.29	48.27
23	27.14	28.43	29.98	32.01	35.17	41.64	49.73
24	28.24	29.55	31.13	33.20	36.42	42.98	51.18
25	29.34	30.68	32.28	34.38	37.65	44.31	52.62
26	30.43	31.79	33.43	35.56	38.89	45.64	54.05
27	31.53	32.91	34.57	36.74	40.11	46.96	55.48
28	32.62	34.03	35.71	37.92	41.34	48.28	56.89
29	33.71	35.14	36.85	39.09	42.56	49.59	58.30
30	34.80	36.25	37.99	40.26	43.77	50.89	59.70
31	35.89	37.36	39.12	41.42	44.99	52.19	61.10
32	36.97	38.47	40.26	42.58	46.19	53.49	62.49
33	38.06	39.57	41.39	43.75	47.40	54.78	63.87
34	39.14	40.68	42.51	44.90	48.60	56.06	65.25
35	40.22	41.78	43.64	46.06	49.80	57.34	66.62
36	41.30	42.88	44.76	47.21	51.00	58.62	67.99
37	42.38	43.98	45.89	48.36	52.19	59.89	69.35
38	43.46	45.08	47.01	49.51	53.38	61.16	70.70
39	44.54	46.17	48.13	50.66	54.57	62.43	72.05
40	45.62	47.27	49.24	51.81	55.76	63.69	73.40
41	46.69	48.36	50.36	52.95	56.94	64.95	74.74
42	47.77	49.46	51.47	54.09	58.12	66.21	76.08
43	48.84	50.55	52.59	55.23	59.30	67.46	77.42
44	49.91	51.64	53.70	56.37	60.48	68.71	78.75
45	50.98	52.73	54.81	57.51	61.66	69.96	80.08
46	52.06	53.82	55.92	58.64	62.83	71.20	81.40
47	53.13	54.91	57.03	59.77	64.00	72.44	82.72
48	54.20	55.99	58.14	60.91	65.17	73.68	84.04
49	55.27	57.08	59.24	62.04	66.34	74.92	85.35
50	56.33	58.16	60.35	63.17	67.50	76.15	86.66

Table A.3. Table shows RIGHT TAIL χ^2 statistics

A.4 The F Distribution

The F distribution is the distribution for a variable that is the ratio of two chi-square distributed random variables. As such, the distribution is nonnegative, just as the chi-square distribution is. Also, the F distribution has two sets of degrees of freedom: "numerator" and "denominator." The table shows the critical F needed to obtain a specified α for a given combination of degrees of freedom.

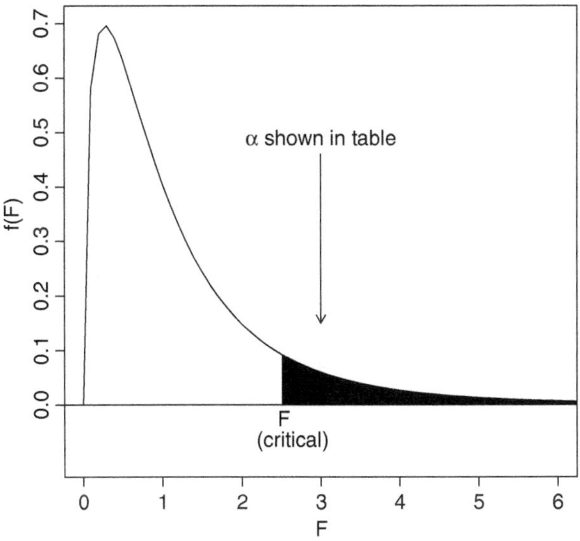

Fig. A.4. The F distribution with 3 and 10 degrees of freedom

($\alpha = .10$)

denom. df.	numerator df									
	1	2	3	4	5	6	7	8	9	10
30	2.88	2.49	2.28	2.14	2.05	1.98	1.93	1.88	1.85	1.82
50	2.81	2.41	2.20	2.06	1.97	1.90	1.84	1.80	1.76	1.73
100	2.76	2.36	2.14	2.00	1.91	1.83	1.78	1.73	1.69	1.66
500	2.72	2.31	2.09	1.96	1.86	1.79	1.73	1.68	1.64	1.61
1000	2.71	2.31	2.09	1.95	1.85	1.78	1.72	1.68	1.64	1.61
1500	2.71	2.31	2.09	1.95	1.85	1.78	1.72	1.67	1.64	1.60
2000	2.71	2.31	2.09	1.95	1.85	1.78	1.72	1.67	1.63	1.60
3000	2.71	2.30	2.09	1.95	1.85	1.78	1.72	1.67	1.63	1.60
5000	2.71	2.30	2.08	1.95	1.85	1.78	1.72	1.67	1.63	1.60
10000	2.71	2.30	2.08	1.95	1.85	1.77	1.72	1.67	1.63	1.60

($\alpha = .05$)

30	4.17	3.32	2.92	2.69	2.53	2.42	2.33	2.27	2.21	2.16
50	4.03	3.18	2.79	2.56	2.40	2.29	2.20	2.13	2.07	2.03
100	3.94	3.09	2.70	2.46	2.31	2.19	2.10	2.03	1.97	1.93
500	3.86	3.01	2.62	2.39	2.23	2.12	2.03	1.96	1.90	1.85
1000	3.85	3.00	2.61	2.38	2.22	2.11	2.02	1.95	1.89	1.84
1500	3.85	3.00	2.61	2.38	2.22	2.10	2.02	1.94	1.89	1.84
2000	3.85	3.00	2.61	2.38	2.22	2.10	2.01	1.94	1.88	1.84
3000	3.84	3.00	2.61	2.37	2.22	2.10	2.01	1.94	1.88	1.83
5000	3.84	3.00	2.61	2.37	2.22	2.10	2.01	1.94	1.88	1.83
10000	3.84	3.00	2.61	2.37	2.21	2.10	2.01	1.94	1.88	1.83

(Continued)

	($\alpha = .01$)									
30	7.56	5.39	4.51	4.02	3.70	3.47	3.30	3.17	3.07	2.98
50	7.17	5.06	4.20	3.72	3.41	3.19	3.02	2.89	2.78	2.70
100	6.90	4.82	3.98	3.51	3.21	2.99	2.82	2.69	2.59	2.50
500	6.69	4.65	3.82	3.36	3.05	2.84	2.68	2.55	2.44	2.36
1000	6.66	4.63	3.80	3.34	3.04	2.82	2.66	2.53	2.43	2.34
1500	6.65	4.62	3.79	3.33	3.03	2.81	2.65	2.52	2.42	2.33
2000	6.65	4.62	3.79	3.33	3.03	2.81	2.65	2.52	2.42	2.33
3000	6.64	4.61	3.79	3.33	3.02	2.81	2.65	2.52	2.41	2.33
5000	6.64	4.61	3.79	3.32	3.02	2.81	2.64	2.51	2.41	2.32
10000	6.64	4.61	3.78	3.32	3.02	2.80	2.64	2.51	2.41	2.32
	($\alpha = .001$)									
30	13.29	8.77	7.05	6.12	5.53	5.12	4.82	4.58	4.39	4.24
50	12.22	7.96	6.34	5.46	4.90	4.51	4.22	4.00	3.82	3.67
100	11.50	7.41	5.86	5.02	4.48	4.11	3.83	3.61	3.44	3.30
500	10.96	7.00	5.51	4.69	4.18	3.81	3.54	3.33	3.16	3.02
1000	10.89	6.96	5.46	4.65	4.14	3.78	3.51	3.30	3.13	2.99
1500	10.87	6.94	5.45	4.64	4.13	3.77	3.50	3.29	3.12	2.98
2000	10.86	6.93	5.44	4.64	4.12	3.76	3.49	3.28	3.11	2.97
3000	10.85	6.92	5.44	4.63	4.11	3.75	3.49	3.28	3.11	2.97
5000	10.84	6.92	5.43	4.62	4.11	3.75	3.48	3.27	3.10	2.97
10000	10.83	6.91	5.43	4.62	4.11	3.75	3.48	3.27	3.10	2.96

Table A.4. Table shows critical values of F for one-tail probabilities

A.5 Using the Tables and Interpolation

The z table follows a different format from the other three tables. In the z table, one finds the first two digits (one to the left of the decimal and one to the right) in the first column and the third digit in the first row and then locates the area of the curve in the table associated with that three-digit z score. In the other tables, one chooses an "alpha" (α) value—an area under the curve—and then finds the "critical" value of the test statistic needed to yield this particular area under the curve. Put another way, in the z table, you use the z to find a p-value, but in the other tables, you use a predefined α (p-value) to find the test statistic that produces it (Table A.4).

The sole reason for the difference in approach using the z versus the other tables is simply the number of parameters involved and constraints on pages. The standard normal distribution has no changing parameters ($\mu = 0$ and $\sigma = 1$). In contrast, the t distribution has a degree of freedom parameter that can take values from 1 to ∞ (or at least 1 to 120). That makes a complete t table 120 times as large as the z table. The χ^2 distribution also has a degree of freedom parameter, and unlike the t distribution, which converges on the normal distribution at 120 d.f., the χ^2 distribution does not converge on the normal distribution so quickly. The F distribution is even larger, because it has two degree of freedom parameters.

It is not uncommon to need a value that is not shown in the t, χ^2 or F tables. For example, suppose one had a t statistic based on a sample with 35 degrees of freedom. The critical t value for a typical two-tailed α of .05 (one-tail $\alpha = .025$) falls between 2.04 and 2.01 (30 d.f. vs. 50 d.f., respectively). So, what if your t statistic were 2.02?

One solution in this type of problem would be to use a program like Excel to compute the exact value for you. An alternative would be to *interpolate* between the values presented in the table to find an approximate value. There are a number of ways one can interpolate between points in the tables, but the basic idea of *linear* interpolation is straightforward: We want to find a value of t for 35 degrees of freedom, given what we know about the values for 30 and 50 degrees of freedom, and we assume the pattern between the points is linear.

To interpolate, consider the degrees of freedom measures as x values, the corresponding t values as y values, and compute the slope $m = \Delta y / \Delta x$. This shows the change in t per unit change in the degrees of freedom; here: $m = (2.01 - 2.04)/(50 - 30) = -.0015$. This means every one unit increase in degrees of freedom from 30 reduces the t statistic by .0015 units from 2.04. Thus, at 35 d.f., $t = 2.04 - 5(.0015) = 2.0325$. In fact, the true value is 2.0301.

Linear interpolation applied to probability distributions will only be approximate, because the distributions we commonly use are not linear: they are exponentially declining in the tails. However, for most purposes, interpolation will be sufficient. It may also be used to approximate the p-value (between given α values).

Appendix B
Answers to Selected Exercises

B.1 Chapter 1

No exercises

B.2 Chapter 2

1. The lay definition, or conceptualization, of a theory connotes a potentially off-the-cuff conjecture or speculation. A scientific theory, however, is a well-supported explanation for how the world works that has stood the test of time; it is not simply speculation.
2. The process of deriving hypotheses from theory is a deductive one. Finding support in data for a hypothesis therefore does not "prove" the theory from which it was drawn. Other explanations may account for the data. Therefore, claiming that the data prove the hypothesis true would be perpetrating a fallacy of affirming the consequent.
3. Perspectives are far too broad to be tested. They are almost always able to posit explanations for any event either via circular reasoning/question-begging or via assuming other facts that are untested or untestable.
4. A research question is a question; a hypothesis is a statement. Usually, hypotheses are written at a level that specifies relationships between actual variables, while a research question may remain at a more abstract level.
5. "Falsifiable" refers to whether it is possible to disprove a hypothesis. For example, the hypothesis that "ghosts exist" is not falsifiable, because we cannot find any kind of evidence to disprove it. In this case, the lack of evidence supporting the hypothesis is not sufficient evidence that ghosts do not exist—we may have simply not gathered enough evidence.

S.M. Lynch, *Using Statistics in Social Research: A Concise Approach*,
DOI 10.1007/978-1-4614-8573-5, © Springer Science+Business Media New York 2013

6. One example: If Sam is lazy, he will be on welfare. Sam is on welfare. Therefore, Sam is lazy. This is fallacious because there may be other explanations for being on welfare, like being disabled or laid-off from one's job.
7. One example: If Sam is lazy, he will be on welfare. Sam is not on welfare. Therefore, Sam is not lazy. Although this argument may seem to be unreasonable from the perspective that the real welfare system does not have a perfect enrollment mechanism (so that, in fact, not all lazy people are successfully enrolled), the formal logic of this argument is solid. The premise states the system of enrollment is perfect.
8. *Modus tollens* enables us to reject/disprove hypotheses. So, bad hypotheses and theories can be ruled out as explanations for how the world works.

B.3 Chapter 3

1. The outcome categories are not mutually exclusive; the end points are included in each category.
2. Ordinal
3. Ratio
4. The individual (micro)
5. It commits an individualistic fallacy. Although it is true that countries with the highest average levels of schooling also have the best health on average, the fact that this pattern exists at the individual level does not imply that it exists at the macro level.
6. There is nothing wrong with this statement. So long as the distribution of IQ is relatively smooth and not terribly skewed (definition discussed in next chapter), the statement is true.
7. It commits an ecological fallacy. In fact, wealthier individuals tend to be Republicans.
8. It is an invalid measure of what we're trying to measure. It is possible for people to support a health care reform bill without a public option, even though they may support a public option. It is also possible to oppose a reform bill whether there is a public option in it or not.
9. Stratified sampling
10. The following items should be asked: age, sex, education, earnings, and hours worked per week. The questions could be asked as follows:

    (a) How old are you in years?
    (b) What is your sex (M/F)?
    (c) How many years of schooling have you completed?
    (d) What are your annual earnings from work?
    (e) How many hours per week do you work on average?

B.4 Chapter 4

1. Histogram of health

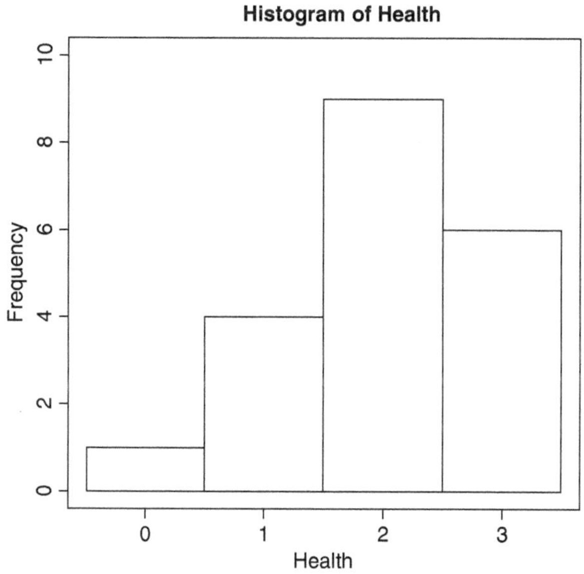

2. Stem and leaf plot of education:

0	9	9					
1	0	0	0				
1	1	1					
1	2	2	2	2	2	2	2
1	3	3					
1	4	4					
1							
1	6	6					

3. Boxplot of education

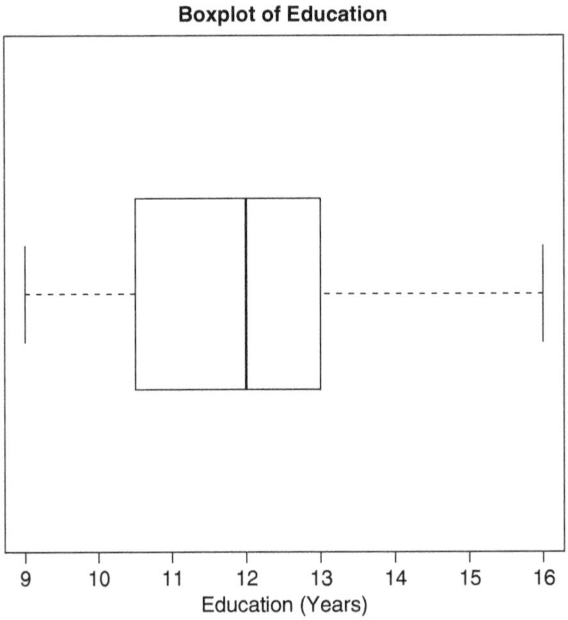

4. $\bar{x} = 12$, MD = 12, MO = 12
5. It appears to be slightly skewed to the right (despite the mean and median being equal, the distribution has a longer right tail than left tail)
6. $\bar{x} = 43.6$, MD = 40.5, MO = 41
7. Range=7, $Q_1 = 10.5$ and $Q_3 = 13$, so IQR=2.5. $s^2 = 3.89$; $s = 1.97$
8. Range=67, $Q_1 = 29.5$ and $Q_3 = 58$, so IQR=28.5. $s^2 = 391.6$; $s = 19.8$
9. Health scores were collapsed into "good" (2,3) and "bad" (0,1). Seventy-five percent of the sample has good health, while 25 % has bad health. This proportion holds for both sexes. Among males, 75 % have good health; among females, 75 % also have good health. There appear to be no sex differences in health in our small sample (note that row percentages were used).

	Health		
Sex	Good	Bad	Total
Male	3	1	4
	75 %	25 %	100 %
Female	12	4	16
	75 %	25 %	100 %
Total	15	5	20
	75 %	25 %	100 %

10. $\overline{health}_{young} = 2.4$, $s_{young} = .70$, $\overline{health}_{old} = 1.6$, $s_{old} = .84$. Younger persons are healthier on average, but the distributions seem to overlap considerably, given that each mean is within roughly one standard deviation of the other.

B.5 Chapter 5

1. The sample space for two children is: $S = \{BB, BG, GB, GG\}$. Only one event represents two girls, and so the probability is $1/4$.
2. If we know that one child is a girl (but not which one), then the original sample space is reduced to: $S = \{BG, GB, GG\}$. So, the probability that both are girls is $1/3$. An alternate way of reaching the conclusion is to use Bayes Theorem. Let B be the event that both children are girls, and let $p(B|one)$ be the probability that both children are girls, given that one is a girl (this is the quantity we want). Then, Bayes rule is:

$$p(B|one) = \frac{p(one|B)p(B)}{p(one|B)p(B) + p(one|\neg B)p(\neg B)} \tag{B.1}$$

$$= \frac{(1)(1/4)}{(1)(1/4) + (2/3)(3/4)} \tag{B.2}$$

$$= \frac{1/4}{1/4 + 1/2} \tag{B.3}$$

$$= \frac{1/4}{3/4} \tag{B.4}$$

$$= 1/3 \tag{B.5}$$

The numerator of the fraction in the first equation the probability that one child is a girl, given that both are girls (1) multiplied by the original (prior) probability that both children are girls (1/4; based on the original sample space). The denominator contains the numerator plus the probability that one child is a girl given that both children are *not* girls multiplied by the initial probability that both children are not girls. The latter term in this product is the complement of the probability that both children are girls (3/4). The former term is the probability that one child is a girl if we know they both are not. This is $2/3$, because if we know both are not girls, there are three possible outcomes for two children, and two of them contain one girl.
3. There are 26 letters in the alphabet. If they can repeat, then there are $(26)(26)(26) = 26^3 = 17{,}576$ possible sequences of the 3 letters. There are 10 digits between 0 and 9. If they can repeat, then there are $(10)(10)(10) = 10^3 = 1{,}000$ possible sequences of digits. Each sequence of letters can be paired with each sequence of digits, so there are $26^3 \times 10^3 = 17{,}576{,}000$ unique plates that can be produced.
4. If the letters and digits cannot repeat, then there are $P(26, 3) = \frac{n!}{(n-x)!} = \frac{26!}{(26-3)!} = (26)(25)(24) = 15{,}600$ possible sequences of letters. Similarly, there are $(10)(9)(8) = 720$ possible sequences of numbers. Thus, there are $(15{,}600)(720) = 11{,}232{,}000$ possible plates that can be produced.

5. We need to determine the ratio of successful events to the size of the sample space. There is only one way to spell "John," and so 1 is the numerator of the ratio. There are $26^4 = 456{,}976$ possible sequences of repeatable letters. Thus, the probability of obtaining a plate with your name on it is $1/(456{,}976)$.
6. This problem is a law of total probability problem. There are two states of the world: the state in which you roll the 6-sided die ($d6$) and the state in which you roll the 10-sided die ($d10$). Within (conditional on) those states of the world, there is a probability of rolling a 5 ($r5$). The roll of the 4-sided die determines which state the world will be in. So:

$$
\begin{aligned}
p(r5) &= p(r5|d6)p(d6) + p(r5|d10)p(d10) && \text{(B.6)}\\
&= \left(\frac{1}{6}\right)\left(\frac{3}{4}\right) + \left(\frac{1}{10}\right)\left(\frac{1}{4}\right) && \text{(B.7)}\\
&= \frac{3}{24} + \frac{1}{40} && \text{(B.8)}\\
&= \frac{3}{20} && \text{(B.9)}
\end{aligned}
$$

The first half of the equation above shows that the probability of rolling a 5 on the six-sided die is 1/6, and this is multiplied by the probability of rolling the six-sided die. That probability is 3/4, the probability of rolling a 1, 2, or 3 on the four-sided die. The latter half of the equation show the probability of rolling a 5 on the 10-sided die (1/10), and this probability is multiplied by the probability of rolling the 10-sided died. That probability is 1/4, the probability of rolling a 4 on the four-sided die.
7. This problem is a Bayes' Theorem problem. We want the probability that the world is/was in state $d10$ (you rolled the 10-sided die) given that you know $r5$ (you rolled a 5). This is $p(d10|r5)$. Given the results presented in the previous problem, we know $p(r5|d10) = 1/10$, $p(d10) = 1/4$, $p(r5|d6) = 1/6$, and $p(d6) = 3/4$. So:

$$
\begin{aligned}
p(d10|r5) &= \frac{p(r5|d10)p(d10)}{p(r5|d10)p(d10) + p(r5|d6)p(d6)} && \text{(B.10)}\\
&= \frac{1/40}{3/20} && \text{(B.11)}\\
&= \frac{1}{6} && \text{(B.12)}
\end{aligned}
$$

Consider this solution for a moment. Without any information, the prior probability of rolling the 10-sided die was 1/4. After observing that a five was rolled, the posterior probability that the 10-sided die was rolled decreases. It does so, because the 10-sided die has more options for outcomes than the six-sided die. Given that we now know one of the outcomes that is duplicated on both dice was rolled, it increases the chance that the smaller die was rolled.

8. 130 is 2 standard deviations above the mean, so there is a 2.5 % chance of obtaining such a person. Technically, 2.5 % of the mass of a normal distribution is above 1.96 standard deviations, but we often use the rule of 2.
9. Selecting two people at random implies the two are independent. So, $p(x_1 > 130, x_2 > 130) = p(x_1 > 130) \times p(x_2 > 130) = (.025)(.025) = .000625$.
10. There are $\binom{6}{3} = 20$ possible samples. Obviously, there can be no samples of size $n = 3$ that contain 0 red marbles. There are 4 samples that contain only 1 red marble: those samples contain both green marbles and one of the four red marbles. There are $\binom{4}{2} \times \binom{2}{1} = 12$ samples that contain 2 red marbles, and there are 4 samples that contain 3 red marbles ($\binom{4}{3} \times \binom{2}{0}$). So, the height of the histogram bars at frequencies 0, 1, 2, and 3 should be 0, 4, 12, and 4.

    We discussed this type of computation in Chap. 3 in describing the importance of random sampling. While we have worked this type of sampling process out logically, the mass function for the hypergeometric distribution formalizes it. The mass function is:

$$p(x, n - x) = \frac{\binom{X}{x}\binom{N-X}{n-x}}{\binom{N}{n}}. \tag{B.13}$$

    The hypergeometric distribution tells us the probability of obtaining x objects of one type and $n - x$ objects of another type in a sample of size n drawn *without replacement* from a population in which there are N total objects, of which X are of one type and $N - X$ are of another. The denominator represents the number of ways n items can be selected from N total objects (like drawing 3 marbles out of 6 total marbles). The numerator represents the number of ways x items can be taken from a total of X items (like drawing 2 red marbles from 4 red ones) multiplied by the number of ways $n - x$ items can be selected from $N - X$ items (like drawing 1 green marble from 2 green marbles).
11. $p(HHHHT) = \frac{1}{2} \times \frac{1}{2} \times \frac{1}{2} \times \frac{1}{2} \times \frac{1}{2} = \left(\frac{1}{2}\right)^5 = \frac{1}{32}$. Notice that the order of heads and tails is specified in the question.
12. We could take the above answer and realize that the one tail could come first, second, third, fourth, or fifth, so there are 5/32 ways to get 4 heads and 1 tail. Otherwise, we can use the binomial formula:

$$p(x = 4) = \binom{5}{4} .5^4 \times .5^1 \tag{B.14}$$

    Notice that the latter product is the same as in the previous problem. The only difference here is the combinatorial, which simply quantifies the number of ways in which the tail can be distributed among the heads.
13. Use the binomial mass function:

$$p(x = 4) = \binom{5}{4} .8^4 \times .2^1 = .4096. \tag{B.15}$$

14. There are a total of $6^5 = 7{,}776$ possible rolls of a set of 5 dice (6 possibilities on the first die, 6 on the second, and so on). Six of the possible are yahtzees (all ones, twos, threes…). So, the probability is $6/7{,}776 = 1/1{,}296 = .000772$
15. To answer this question, we must compute two z scores (standardized scores) and then find the appropriate area between.

$$z_1 = \frac{175 - 190}{59} = -.254 \tag{B.16}$$

$$z_1 = \frac{200 - 190}{59} = .169 \tag{B.17}$$

16.7 % fall in this range.
16. If the men are independently sampled, then the joint probability is the product of their respective probabilities. So: $p = .167^5 = .00013$
17. If 20 % of males are obese, then 20 % of 50 % (10 %) are in the overlap region of "male" and "obese." This implies the same Venn diagram as in the chapter.
18. $p(male \quad or \quad obese) = p(male) + p(obese) - p(male, obese) = .5 + .3 - .1 = .7$
19. $p(obese, male) = .1$
20. $p(male|obese) = \frac{p(obese,male)}{p(obese)} = \frac{.1}{.3} = .333.$
21. This is a straighforward standardized score problem. Once we convert height into the z metric, we can find the proportion of the distribution that falls below z:

$$z = \frac{68 - 71}{4} = -.75. \tag{B.18}$$

$p(z < -.75) = .227$. So, given that roughly 22–23 % of the distribution is shorter, I would be roughly in the 23rd percentile of height.
22. I could apply the binomial mass function repeatedly and sum up the probability of obtaining 55 heads, 56 heads, 57 heads, and so on, up to 100 heads. If I did so, I would obtain a probability of .184. Alternatively, I could use the normal approximation to the binomial. First, the probability of obtaining 55 or more heads is equivalent to 1 minus the probability of obtaining 54 heads or fewer. Thus, we compute:

$$z = \frac{(x + .5) - np}{\sqrt{np(1-p)}} \tag{B.19}$$

$$= \frac{54.5 - (100)(.5)}{\sqrt{(100)(.5)(.5)}} \tag{B.20}$$

$$= .9 \tag{B.21}$$

The z table in the appendix only shows probabilities for *negative* valus of z, and this makes the problem somewhat tedious, because we have to think about what region under the curve we are ultimately interested in. The z score divides the normal distribution into two pieces. Under the symmetry of the normal distribution, $p(z > .9) = p(-z < -.9) = .184$. Recall that we are interested in the probability that the we obtain 55 or more heads. This is the probability that $p(z > .9)$ under our approximation, and that is equal to $p(-z < -.9)$, which we found. Thus, we do not need to subtract from 1 after all. The probability is .184.

Thinking about this problem slowly and carefully helps us reach this decision. With a fair coin, we would expect 50 heads ($np = 100 \times .5$). Obtaining more heads than this should lead us to expect that the probability associated with obtaining more than the expected number of heads should be less than .5. .184 is less than .5. If we subtracted .184 from 1, we would obtain .816, which is much greater than .5. This result, if nothing else, should convince us that the subtraction is not needed.

23. The order in which the cards are dealt is irrelevant. So: $\binom{52}{5} = 2{,}598{,}960$
24. There are four royal flushes (one for each suit). So: $p(r.f.) = 4/2{,}598{,}960$
25. There are at least two ways to consider this problem. Imagine taking the top five cards off the deck. The probability that the top five cards are of matching suit is the probability that the second card's suit matches the first, the third card's suit matches the second, and so on. So:

$$p(f) = \left(\frac{12}{51}\right)\left(\frac{11}{50}\right)\left(\frac{10}{49}\right)\left(\frac{9}{48}\right) = .001981 \tag{B.22}$$

A second way to solve this problem is to compute the total number of flushes that exist out of the total number of poker hands. Within each suit, there are $\binom{13}{5} = 1{,}287$ unique flushes. There are four suits. So:

$$p(f) = \frac{4 \times 1{,}287}{2{,}598{,}960} = .001981 \tag{B.23}$$

B.6 Chapter 6

1. In this problem, μ and σ are known, so the probability can be assessed using a z statistic:

$$z = \frac{11 - 12}{3/\sqrt{50}} = -2.36. \tag{B.24}$$

$p(z < -2.36) = .009$. This is the probability of obtaining such a sample with a mean less than 2.36 standard errors below the mean. Thus, the probability of obtaining a sample with a mean greater than 11 is .991.

2. Again, in this problem, μ and σ are known, so use z:

$$z = \frac{12.1 - 12}{3/\sqrt{100}} = .33. \tag{B.25}$$

The question asks for the area to the left of 12.1, so $p(z < .33)$. The table does not give us this probability directly, but realize that the area to the left of $z = 0$ is .5. We need to add to that the area between $z = 0$ and $z = .33$. By symmetry, $p(z > .33) = p(-z < -.33) = .371$, and the area between $z = 0$ and $z = .33$ is $.5 - .371 = .129$. Thus, the probability of obtaining a sample with a mean less than 12.1 is $.50 + .129 = .629$.
3. Again, μ and σ are known, so use z:

$$z = \frac{11.8 - 12}{3/\sqrt{500}} = -1.49. \tag{B.26}$$

$p(z < -1.49) = .068$. Realize that 11.8 and 12.2 are equidistant from the mean, and so the probability of obtaining a sample with a mean below 11.8 is the same as the probability of obtaining one with a mean above 12.2. So, the answer is $(.068)2 = .136$.
4. There are two ways to answer this question: using the z statistic for proportions and using the binomial distribution. So:

$$z = \frac{\hat{p} - p}{\sqrt{p(1-p)/n}} = \frac{.5 - .52}{\sqrt{.52(.48)/200}} = -.566. \tag{B.27}$$

The probability that z is below that is .286 (based on interpolating results from the z table between $z = -.56$ and $z = -.57$).

Alternatively, obtaining a sample of 200 in which less than half supported a candidate when 52 % in the population do implies summing binomial probabilities for 0–99 "successes," when the binomial distribution has parameters $n = 200$ and $p = .52$:

$$\sum_{x=0}^{99} \binom{200}{x} .52^x .48^{200-x} = .262. \tag{B.28}$$

The difference between the answer using the z distribution and the binomial distribution is that the latter is discrete: there is no person between the 99th and 100th. The normal distribution, however, is continuous, and so "less than half" includes all real numbers from 99 up to (but not including) 100. In fact, there is a probability of .048011 that exactly 100 persons would support the candidate. If we add half of this probability to the probability of obtaining 99 or fewer supporters, we obtain .286. This process essentially illustrates—in reverse fashion—the normal approximation to the binomial distribution.

5. The null hypothesis is that $\mu = 120$. The appropriate test is the one sample z test, because σ is known (or assumed known):

$$z = \frac{130 - 120}{10/\sqrt{50}} = 7.07. \tag{B.29}$$

The probability that we could obtain such a sample mean if this null hypothesis were true is $p(|z| > 7.07) \approx 0$, and so we would reject our hypothesis that $\mu = 120$.

6. Construct a confidence interval. Given that $n > 120$, $t = 1.96$ for a 95 % interval:

$$13.58 \pm (1.96)(2.66/\sqrt{2386}) = [13.47, 13.69] \tag{B.30}$$

I am 95 % confident that the true population mean is between 13.47 and 13.69.

7. The null hypothesis is that $p_0 = .333$. An appropriate test is a one sample z test for proportions:

$$z = \frac{.324 - .333}{\sqrt{.333(.667)/26228}} = -3.09. \tag{B.31}$$

$p(|z| > 3.09) < .01$, so we would reject the null hypothesis and conclude that one-third of the population is not in excellent health. Notice, however, that this "statistically significant" result is probably not particularly interesting: .324 *is* approximately one-third to most people.

8. The null hypothesis is that happiness levels are equal across income groups. A proper test is an independent samples t test:

$$t = \frac{1.32 - 1.11}{\sqrt{.59^2/11366 + .64^2/14862}} = 27.53 \tag{B.32}$$

The sample is large enough that df=∞, so $p(|t| > 27.53) \approx 0$. Reject null and conclude that happiness levels differ by income group. Substantively, we would conclude that income does predict happiness.

9. For the high income group: $1.32 \pm (1.96)(.59/\sqrt{11366}) = [1.31, 1.33]$
For the low income group: $1.11 \pm (1.96)(.64/\sqrt{14862}) = [1.10, 1.12]$
These intervals do not overlap, providing another way to conclude that happiness levels differ by income group.

10. A proper test is an independent samples t test with a null hypothesis that means do not differ:

$$t = \frac{23.03 - 20.55}{\sqrt{4.52^2/10791 + 5.07^2/1269}} = 16.66. \tag{B.33}$$

The degrees of freedom exceed 120, so $p(|t| > 16.66) < .001$. Thus, we should reject the null and conclude that mean satisfaction varies between blacks and whites.

11. A proper test is an independent samples t test with a null hypothesis that means do not differ:

$$t = \frac{53641 - 54291}{\sqrt{31213^2/1106 + 40511^2/1420}} = -.46. \tag{B.34}$$

$p(|t| > .46) > .05$, so, we cannot reject null that mean wages were the same in '73 and '06.

12. The question can be posed as whether two independent proportions differ; a proper test is an independent samples t test for proportions with a null hypothesis that there is no difference between sexes:

$$t = \frac{.7 - .65}{\sqrt{.7(.3)/10 + .65(.35)/8}} = .22 \tag{B.35}$$

The degrees of freedom for assessing this t statistic is 7 (one less than the minimum subsample size). We cannot reject null that men's and women's tastes are the same.

13. The null hypothesis in this case is not 0: it is 3.5; the question is whether the population mean could be 3.5, given a sample mean of 3.32. So:

$$t = \frac{3.32 - 3.5}{1.71/\sqrt{1420}} = -3.97. \tag{B.36}$$

Given the sample size, the df are 120+, so $p(|t| > 3.97) < .001$. We should therefore reject the null hypothesis and conclude that population isn't in the middle of the scale. As in an earlier problem, however, this finding may not be substantively meaningful, given that 3.32 *is* close to 3.5.

14. The question is whether the mean ages are different in '72 and '06. A proper test is an independent samples t test with a null hypothesis that they are not different. So:

$$t = \frac{43.89 - 46.10}{\sqrt{16.88^2/1086 + 16.85^2/1420}} = -3.25. \tag{B.37}$$

The degrees of freedom here are "∞," so $p(|t| > 3.25) < .001$. Thus, we should reject null hypothesis that mean ages are the same across the time period and conclude that ages are different. In other words, yes, the GSS is consistent with the view that the population is aging. (Note: it should be, because there is no question that the population is aging!)

15. While an independent samples t test might seem appropriate for this question, that test is not an efficient method for answering the question, because married

couples are not independent individuals. Instead, a paired sample t test is appropriate. For that test, we compute the pair-wise differences in IQ and then produce a one-sample t statistic from the mean and standard deviation of the paired differences. The table below shows the differences. The mean difference is 0, and the standard deviation is about 16.

Couple	Husband	Wife	diff
1	100	110	−10
2	140	135	5
3	90	120	−30
4	150	120	30
5	140	130	10
6	95	110	−15
7	100	95	5
8	50	50	0
9	200	200	0
10	100	95	5
			$\bar{x}_{diff} = 0$
			$s_{diff} = 15.99$

The null hypothesis is that $\mu_{diff} = 0$. So:

$$t = \frac{0}{15.99/\sqrt{10}} = 0. \tag{B.38}$$

$p(|t| > 0) = 1$ so we can't reject null. Husbands' and wives' intelligence is comparable.

Note that both the independent samples t test and the paired sample t test calculations will produce the same numerator: the mean of a set of differences between two groups is equal to the difference in the means of the two groups. It is the denominator that differs, with the denominator of the paired sample t test generally being much smaller. In this problem, if men and women have roughly the same mean IQ in the population, then you might suspect that paired differences for married couples would also have a mean of 0. However, this assumes that the process of marriage is a random one, and it isn't: marriage is selective. If there were a tendency for, say, more intelligent men to marry less intelligent women, then highly intelligent women would remain unmarried, and our paired sample of married couples would reflect the difference.

16. The null hypothesis is that men and women's incomes are the same (i.e., $\mu_{men} - \mu_{women} = 0$). So:

$$t = \frac{47828 - 41354}{\sqrt{(30031)^2/277 + (26333)^2/351}} = 2.83 \tag{B.39}$$

$min(n_m - 1, n_w - 1) > 120$, so we can use $t = z = 1.96$ as our critical t. $p(|t| > 2.83) < .05$ (indeed, $p < .01$), so we can reject the null. Men and women's incomes differ. In particular, it looks as though men's income is greater than women's, and differences in education level cannot account for this difference, because we have "controlled" on educational differences by limiting to men and women with the same level of education. This notion of "controlling" is a key feature of regression analyses, as discussed in subsequent chapters.

17. This question is set-up for a chi-square test of independence, as is discussed in the next chapter. However, it can also be answered using an independent samples t-test, so long as we are willing to make the assumption that there is equal spacing between the categories in the "liberal to conservative" measure. (Recall that this is the key distinction between ordinal and interval/continuous measures). If we make that assumption, then it does not matter what numbers we assign to the categories, so we may as well assign 1,2,3 to the "liberal," "moderate," and "conservative" outcomes. If we do this, we can compute the means and variances for each group as follows:

Sex	category	code	n	$n \times x$	$\bar{x}$
Female	liberal	1	487	487	
Female	moderate	2	448	896	
Female	conservative	3	366	1098	1.91
Male	liberal	1	332	332	
Male	moderate	2	416	832	
Male	conservative	3	337	1011	2.00

Variances can be computed as $\sum n_c(x - \bar{x}_c)^2/n$ and are .65 and .62, respectively. Then:

$$t = \frac{1.91 - 2.00}{\sqrt{.65/1301 + .62/1085}} = -2.99 \quad \text{(B.40)}$$

Given the sample size, df> 120, so $p(|t| > 2.99) < .01$, and we can reject the null that political views are equivalent. It seems that men are more conservative than women.

18. I'd construct a confidence interval here and see if it overlaps .5. If not, I'd call the election. $n > 120$, so $t = z = 1.96$ at the $\alpha = .05$ level (as in an earlier problem). So:

$$CI = .55 \pm (1.96)(\sqrt{(.55)(.45)/400}) \quad \text{(B.41)}$$

$$= .55 \pm .049 \quad \text{(B.42)}$$

$$= [.501, .599] \quad \text{(B.43)}$$

We are 95 % confident that the proportion voting for candidate A is above .50, so we could call the election in favor of candidate A. Similarly, if we had conducted a t-test with a null hypothesis that $p_0 = .5$, we would reject the null at the $\alpha = .05$ level. HOWEVER, realize that, if we wanted to be more confident about p_0, our interval would widen, and we'd conclude that the election is too close to call.

19. The null hypothesis is that $\mu = 50{,}000$. So:

$$t = \frac{51120.18 - 50000}{32045.84/\sqrt{2386}} = 1.71 \tag{B.44}$$

$p(|t| > 1.71) > .05$, so we cannot reject the null. Our neighbor's estimate is reasonable.

20. The margin of error (MOE) is the latter part of the confidence interval:

$$t_{\alpha/2}\left(s/\sqrt{n}\right). \tag{B.45}$$

Usually, we assume a 95 % confidence level, and because the sample size will certainly be larger than 120, $t = 1.96$. Next, recall that the variance of a proportion is a function of the proportion itself ($p(1-p)$). This variance is maximized at $p = .5$. Thus, if we want to be certain our sample size is large enough to guarantee an MOE of ± 2 %, we should set $p = .5$ and solve for n:

$$.02 = 1.96(.5)(.5)/\sqrt{n} \tag{B.46}$$

$$n = 2401 \tag{B.47}$$

Notice that the result here is a whole number. If it hadn't been, we would need to round up to the next whole person, because we cannot interview a fraction of a person.

B.7 Chapter 7

1. The interpretation should be in terms of the column percentages, e.g.: Only 13.3 % of married folks live in the northeast, while more than one-third of them live in the south (35.9 %); overall, 36 % of the sample lived in the south, yet 41.8 % of those who are divorced/separated live in the south, etc.

Region	Marital Status Married	Widowed	Divorced/ Separated	Never Married	TOTAL
Northeast	91	20	29	61	201
	13.3 %	18.2 %	10.5 %	17.4 %	14.2 %
Midwest	174	30	71	79	354
	25.4 %	27.3 %	25.8 %	22.6 %	24.9 %
South	246	30	115	120	511
	35.9 %	27.3 %	41.8 %	34.3 %	36.0 %
West	174	30	60	90	354
	25.4 %	27.3 %	21.8 %	25.7 %	24.9 %
TOTAL	685	110	275	350	1420
	100 %	100 %	100 %	100 %	100 %

2. A fair die would allocate rolls to each number 1–6 equally. So, the true proportions would be $1/6, 1/6 \ldots 1/6$. In any given sample, however, randomness may lead to higher or lower counts than these proportions suggest. Can the differences in counts be attributed to random fluctuation? An appropriate test is a lack-of-fit test:

Outcome	Observed count	Expected ($1/6 \times 1000$)	$\frac{(O-E)^2}{E}$
1	152	166.7	1.30
2	163	166.7	.08
3	146	166.7	2.57
4	173	166.7	.24
5	171	166.7	.11
6	195	166.7	4.80
Total	1000	1000	$\chi^2 = 9.1$

$\chi^2 = 9.1$ on 5 d.f. 9.1 is less than 11.07, the critical value for $\alpha = .05$; indeed, 9.1 is less than 9.24, the critical value for $\alpha = .1$. So, we cannot reject the hypothesis that these data came from a fair die.

3. Chi square test of independence (cell contributions to the total χ^2 shown in table):

	Happiness			
Party	Unhappy	Somewhat Happy	Very Happy	Total
Democratic	1.0	2.1	6.8	9.9
Independent	1.7	0	1.2	2.9
Republican	5.2	2.8	14.1	22.1
Total	7.8	5.0	22.1	34.9

$\chi^2 = 34.9$ on 4 d.f., $p < .001$ (critical value is 18.47 for $\alpha = .001$ level). Thus, we can reject the null hypothesis that happiness and party are independent and conclude that party and happiness are associated. From the expected counts, it appears that there is an excess of happy republicans, too few happy democrats, too many unhappy democrats, and too few unhappy republicans.

4. Chi square test of independence (cell contributions shown in table):

	Happiness			
Health	Unhappy	Somewhat Happy	Very Happy	Total
Poor	21.9	.1	6.6	28.6
Fair	18.8	.2	11.7	30.8
Good	1.8	3.1	2.4	7.2
Excellent	14.7	6.5	34.8	56.0
Total	57.2	9.9	55.6	$\chi^2 = 122.6$

$\chi^2 = 122.6$ on 6 d.f., $p < .001$ (critical value is 22.46 for $\alpha = .001$ level). Thus, we can reject the null of independence and conclude that health and happiness are associated. In general, it seems that as health increases, so does happiness.

5. An appropriate test is a lack-of-fit test:

Outcome	Observed count	Expected	$\frac{(O-E)^2}{E}$
Male	12,038	12,851.72	51.52
Female	14,190	13,376.28	49.50
Total	26,228	26,228	$\chi^2 = 101.02$

$\chi^2 = 101.02$ on 1 d.f., $p < .001$. Thus, we should reject the hypothesis that this sample came from these population proportions.

B.8 Chapter 8

1. The null hypothesis is that mean income does not vary by race. Based on the table, we can reject the null and say that income varies by race.

Source	Sum of Squares	DF	Mean Squares	F
Between	78169.41	2	39084.71	24.61 ($p < .001$)
Within	2250569.72	1417	1588.26	
Total	2328739.13	1419	1641.11	$R^2 = .034$

2. The null hypothesis is that mean energy level change is the same across treatment groups. Based on the F statistic, we cannot reject this null.

Person	Treatment Group		
	Treatment	Placebo	Control
1	0	1	2
2	3	1	0
3	1	2	0
4	5	3	3
5	2	0	1
$\bar{x} =$	2.2	1.4	1.2
Sum of Squares	14.8	5.2	6.8
$\bar{\bar{x}} = 1.6$			
SST=29.6			

Source	Sum of Squares	DF	Mean Squares	F
Between	2.8	2	1.4	.63 ($p > .1$)
Within	26.8	12	2.23	
Total	29.6	14	2.11	$R^2 = .095$

3. The SST can be obtained by squaring the sample standard deviation and multiplying by one less than the sample size ($n - 1$). The SSW can be obtained by squaring each of the subsample standard deviations, multiplying by one less than the respective subsample sizes, and summing. The remaining calculations can be carried out as in the previous exercises. The F statistic implies a rejection of the null hypothesis, and so, we conclude that mean health varies by education level.

Source	Sum of Squares	DF	Mean Squares	F
Between	49.36	2	24.68	38.56 ($p < .001$)
Within	904.78	1417	.64	
Total	954.14	1419	.67	$R^2 = .052$

B.9 Chapter 9

Below is the data set for the exercises, with additional columns containing key computations for use in answering the questions. The means of each variable are presented. They are used to construct the new columns (1–4), which are the deviations of each value from the means of the respective variable. The variances are found by (1) squaring the values in each new column, (2) summing up these squared deviations (within columns), and (3) dividing the sum by $n - 1 = 9$.

ID	S	He	Ha	E	(1) $(S - \bar{S})$	(2) $(He - \bar{He})$	(3) $(Ha - \bar{Ha})$	(4) $(E - \bar{E})$
1	21	3	2	17	−.9	.9	.7	3.2
2	19	2	1	14	−2.9	−.1	−.3	.2
3	25	2	1	13	3.1	−.1	−.3	−.8
4	26	3	1	16	4.1	.9	−.3	2.2
5	26	2	1	12	4.1	−.1	−.3	−1.8
6	24	3	1	16	2.1	.9	−.3	2.2
7	20	1	3	17	−1.9	−1.1	1.7	3.2
8	10	1	0	8	−11.9	−1.1	−1.3	−5.8
9	23	2	1	13	1.1	−.1	−.3	−.8
10	25	2	2	12	3.1	−.1	.7	−1.8
$\bar{x}$	21.9	2.1	1.3	13.8				
s^2					23.66	.54	.68	7.96

1. Relevant columns are (1) and (3). Multiply the values together by row and sum them to obtain $\sum(x - \bar{x})(y - \bar{y}) = -.63 + .87 + \ldots + 2.17 = 10.3$. Divide this quantity by $n - 1 = 9$ to obtain the covariance ($= 1.1444$). Obtain the correlation by dividing the covariance by the standard deviation of satisfaction (=4.86) and happiness (=.82). Result is: $r = .286$.

   To obtain the confidence interval, use the following steps:

   (a) Compute Fisher's z transformation of r:

   $$z_f = .5 \times [\ln(1 + r) - \ln(1 - r)] = .294 \tag{B.48}$$

   (b) Compute the standard error: $\sigma_{z_f} = 1/\sqrt{n - 3} = .378$.
   (c) Compute the upper and lower bounds as $z_f \pm 1.96 \times \sigma_z = [-.455, 1.03]$ for a 95 % interval.
   (d) Transform the bounds back into the r metric by inverting the z transformation: $L = \frac{\exp(2(-.455))-1}{\exp(2(-.455))+1} = -.42$; $U = \frac{\exp(2(1.03))-1}{\exp(2(1.03))+1} = .77$

   95 % Confidence interval is [−.42, .77]. Given that the interval overlaps 0, we cannot reject the null hypothesis that the true correlation in the population is 0. Therefore, we conclude that life satisfaction and happiness are not related.

   The z test approach yields the same conclusion:

   $$z = \frac{z_f - 0}{\hat{\sigma}_{z_f}} \tag{B.49}$$

   $$= \frac{.294}{.378} \tag{B.50}$$

   $$= .78. \tag{B.51}$$

This z score is less than the 1.96 critical value of z needed to reject the null hypothesis, and so we cannot rule out that $\rho = 0$ in the population.

2. The formula for the slope in the simple regression model is:

$$\hat{\beta} = \frac{\sum(x - \bar{x})(y - \bar{y})}{\sum(x - \bar{x})^2}. \tag{B.52}$$

We can obtain these values from columns (2) and (4). First multiply the values in these columns together and sum them to obtain 10.2. Next, square each item in column (4) and sum these values to obtain 71.6. Then divide 10.2 by 71.6 to obtain $\hat{\beta} = .142$. This result implies that each unit increase in education is associated with an expected increase of .142 units of happiness.

The intercept is easy to compute:

$$\hat{\alpha} = \bar{y} - \hat{\beta}\bar{x} \tag{B.53}$$

$$= 2.1 - .142(13.8) \tag{B.54}$$

$$= .140 \tag{B.55}$$

The prediction equation is therefore: $\hat{y} = .14 + .142 Education$. We use these results to construct the ANOVA table. To do this, we need to perform a series of computations, starting with computing the model predicted values for health for each person in order to obtain the errors (and ultimately the error sums of squares).

Person	E	H	$\hat{H}$	$e = H - \hat{H}$	e^2
1	17	3	2.55	.45	.2025
2	14	2	2.13	−.13	.0169
3	13	2	1.99	.01	.0001
4	16	3	2.41	.59	.3481
5	12	2	1.84	.16	.0256
6	16	3	2.41	.59	.3481
7	17	1	2.55	−1.55	2.4025
8	8	1	1.28	−.28	.0784
9	13	2	1.99	.01	.0001
10	12	2	1.84	.16	.0256
					$\sum e^2 = 3.45$

We can obtain the total sum of squares by using the sum of squared values in column (2); the sum is 4.9. We have enough information now to complete the table:

Source	Sums of Squares	df	Mean Squares	F
Model	1.45	1	1.45	3.36
Error	3.45	8	.431	
Total	4.90	9	.54	

The F statistic on 1 and 8 df is smaller than the value of 4.17, which is the critical value for $\alpha = .05$ on even more (30) degrees of freedom than we have. Therefore we cannot reject the null that there is no linear association between education and health.

In terms of the t-tests on the intercept and slope, we can compute the standard errors for these parameter estimates using primarily values we have already computed. For the intercept, we need to obtain the sum of squares of the original values of education, but the rest of the quantities needed have already been computed. σ_e^2 can be estimated using the MSE of .431.

$$s.e.(\hat{\alpha}) = \sqrt{\frac{\sigma_e^2 \sum x_i^2}{n \sum (x_i - \bar{x})^2}} \tag{B.56}$$

$$= \sqrt{\frac{.431(1976)}{10(71.6)}} \tag{B.57}$$

$$= 1.09 \tag{B.58}$$

$$s.e.(\hat{\beta}) = \sqrt{\frac{\sigma_e^2}{\sum (x_i - \bar{x})^2}} \tag{B.59}$$

$$= \sqrt{\frac{.431}{71.6}} \tag{B.60}$$

$$= .078 \tag{B.61}$$

(B.62)

We usually put the coefficients, standard errors, and t test statistics in a table as follows:

Parameter	Coefficient	S.E.	t	p
Intercept	.14	1.09	.13	$p > .1$
Slope	.142	.078	1.82	$p > .1$

Given that the usual hypothesis is that the coefficients are 0 in the population, the t statistic is simply the ratio of the coefficient to its s.e. We find that, on 8 degrees of freedom (the denominator degrees of freedom—$n - k$, where k is the number of intercept and slope parameters estimated), both t statistics are smaller than 2.31, the critical value needed to reject the null hypothesis. Thus, we can conclude that the effect of education on health in the population is 0, meaning there is no relationship.

B.10 Chapter 10

1. The three models are different polynomial regressions for the shape of the education-income relationship. The first model says that the relationship is linear; the second says that the relationship is quadratic; and the third says the relationship is cubic. The figure below shows all three lines. From the appearance of the lines over regions of the data with lots of data (above 5 years of schooling), the three lines appear to produce similar pictures of the education-income relationship. So, despite the fact that the quadratic and cubic coefficients are statistically significant, a linear specification seems sufficient.

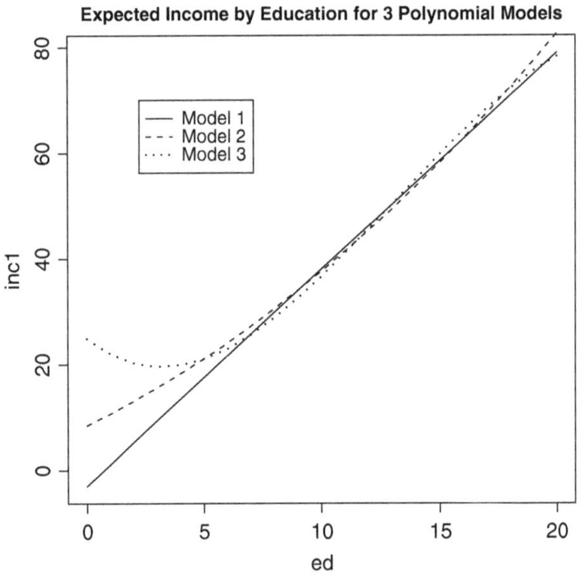

2. The model implies two lines: one for men and one for women. The interaction implies that the lines are not parallel, but instead, converge across education. Men make more than women, but their slope across education is shallower than that for women.

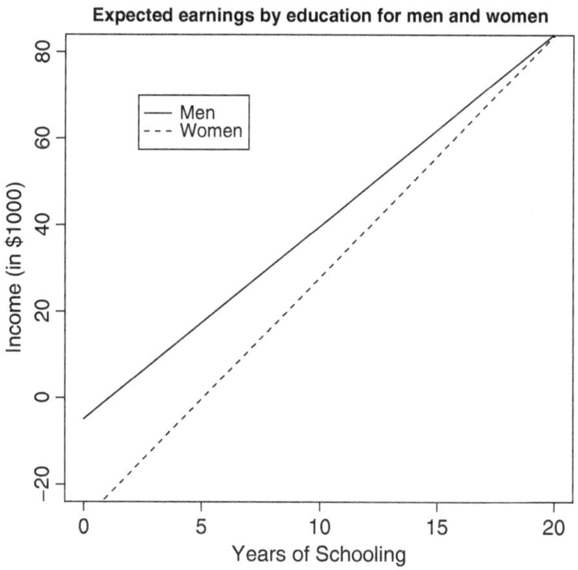

3. The results in the table show that this combination of variables accounts for between one-quarter and one-third of the total variance in income across the sample ($R^2 = .29$), indicating these are important predictors of income. According to the coefficients and their levels of statistical significance, older persons make more money than younger persons, controlling on the other variables in the model. The effect is highly statistically significant, but substantively small (about $90 per year of age). Each successive birth cohort makes more money than the one before it, controlling on other variables in the model, but this effect, while statistically significant, is also small (about $40 more per cohort). Men make about $5,300 more than women, and each year of education produces about $3,600 more in income. Married persons make considerably more than unmarried persons (about $21,700 more). Southerners earn less than persons who live in other regions (about $740 less), but this effect is not statistically significant, indicating that it may be 0 in the population. Finally persons with better health make more money than persons with worse health: each unit increase in health is associated with about $5,270 in expected additional income.

   Note that, although I have not included "controlling on..." in discussing each variable's effect, it is implicit that these effects are net of all other variables in the model. Also note that, although I have used the term "effect," (which is common in the literature), the model does not directly get at causal relationships. Instead, the model shows the differences in the average income for persons at different levels of each covariate.

B.11 Chapter 11

No exercises.

References

Agin, D. (2006). *Junk science: An overdue indictment of government, industry, and faith groups that twist science for their own gain*. New York: Thomas Dunne Books.

Babbie, E. (2004). *The practice of social research* (10th ed.). Belmont: Thomson Wadsworth.

Bonevac, D. (2003). *Deduction: Introductory symbolic logic* (2nd ed.). Malden: Blackwell.

Bunch, B. (1997). *Mathematical fallacies and paradoxes*. Mineola: Dover.

Campbell, S. K. (2004). *Flaws and fallacies in statistical thinking*. Mineoloa: Dover.

Campbell, D. T., & Stanley, J. C. (1963). *Experimental and quasi-experimental designs for research*. Chicago: Rand-McNally.

Cleveland, W. S. (1993). *Visualizing data*. Summit: Hobart Press.

Coyne, J. A. (2009). *Why evolution is true*. New York: Viking.

Davis, J. A. (1985). *The logic of causal order* (Sage University paper series on quantitative applications in the social sciences, series no. 07–55). Beverly Hills: Sage.

DeGroot, M. H., & Schervish, M. J. (2012). *Probability and statistics* (4th ed.). Boston: Addison-Wesley.

Dickens, W. T., & Flynn, J. R. (2006). Black Americans reduce the racial IQ gap: Evidence from standardization samples. *Psychological Science, 16*(10), 913–920.

Dillman, D. A., Smyth, J. D., & Christian, L. M. (2009). *Internet, mail, and mixed-mode surveys: The tailored design method* (3rd ed.). Hoboken: Wiley.

Duneier, M. (2012). Qualitative methods. In G. Ritzer (Ed.), *The Wiley-Blackwell companion to sociology* (1st ed., pp. 73–81). West Sussex: Blackwell.

Durkheim, E. (1997). *The division of labor in society* (L. A. Coser, Trans.). New York: Free Press.

Firebaugh, G. (2008). *Seven rules for social research*. Princeton: Princeton University Press.

Fox, J. (2008). *Applied regression analysis and generlized linear models* (2nd ed.). Thousand Oaks: Sage.

Gujarati, D. N., & Porter, D. C. (2009). *Basic econometrics* (5th ed.). New York: McGraw-Hill.

Hawking, S. (1988). *A brief history of time*. New York: Bantam.

Hooke, R. (1983). *How to tell the liars from the statisticians*. New York: Marcel Dekker.

Huff, D. (1993). *How to lie with statistics*. New York: W.W. Norton.

Idler, E. L., & Benyamini, Y. (1997). Self-rated health and mortality: A review of twenty-seven community studies. *Journal of Health and Social Behavior, 38*(1), 21–37.

Kuhn, T. S. (1962). *The structure of scientific revolutions*. Chicago: University of Chicago Press.

Lieberson, S. (1987). *Making it count: The improvement of social research and theory*. Berkeley: University of California Press.

Lohr, S. L. (1999). *Sampling: Design and analysis*. Pacific Grove: Duxbury Press.

Lynch, S. M. (2007). *Introduction to applied Bayesian statistics and estimation for social scientists*. New York: Springer.

S.M. Lynch, *Using Statistics in Social Research: A Concise Approach*,
DOI 10.1007/978-1-4614-8573-5,

© Springer Science+Business Media New York 2013

Marx, K. (1988). *Economic and philosophic manuscripts of 1844* (M. Milligan, Trans.). Amherst: Prometheus Books.

Merton, R. K. (1968). *Social theory and social structure*. New York: Free Press.

Mlodinow, L. (2008). *The Drunkard's walk: How randomness rules our lives*. New York: Pantheon Books.

Morgan, S. L., & Winship, C. (2007). *Counterfactuals and causal inference: Methods and principles for social research*. New York: Cambridge University Press.

Paulos, J. A. (2001). *Innumeracy: Mathematical illiteracy and its consequences*. New York: Hill & Wang.

Pigliucci, M. (2010). *Nonsense on stilts: How to tell science from bunk*. Chicago: The University of Chicago Press.

Popper, K. S. (1992). *The logic of scientific discovery*. New York: Routledge.

Preston, S. H., Heuveline, P., & Guillot, M. (2001). *Demography: Meauring and modeling population processes*. Oxford: Blackwell.

Scheaffer, R. L., Mendenhall, W., III, Ott, R. L., & Gerow, K. (2012). *Elementary survey sampling* (7th ed.). Boston: Brooks/Cole.

Smith, T. W., Marsden, P., Hout, M., & Kim, J. (2011). *General social surveys 1972–2010* [machine-readable data file]/Principal Investigator, T.W. Smith; Co-Principal Investigator, P.V. Marsden; Co-Principal Investigator, M. Hout; Sponsored by National Science Foundation.–NORC ed.–Chicago: National Opinion Research Center [producer]; Storrs, CT: The Roper Center for Public Opinion Research, University of Connecticut [distributer].

von Hippel, P. T. (2005). Mean, median, and skew: Correcting a textbook rule. *Journal of Statistics Education, 13*(2).

Western, B. (2009). *Punishment and inequality in America*. New York: Russell Sage Foundation.

Wilson, W. J. (1987). *The truly disadvantaged: The inner city, the underclass, and public policy*. Chicago: The University of Chicago Press.

Wimmer, A. (2013). *Waves of war: Nationalism, state formation, and ethnic exclusion in the modern world*. New York: Cambridge University Press.

Ziliak, S. T., & McCloskey, D. N. (2008). *The cult of statistical significance: How the standard error costs us jobs, justice, and lives*. Ann Arbor: University of Michigan Press.